JMP® User's Guide
Version 3.1 of JMP

SAS Institute Inc.
SAS Campus Drive
Cary, NC 27513

JMP® User's Guide, Version 3.1

© Copyright 1995 by SAS Institute Inc., Cary, NC, USA

ISBN 1-55544-678-7

All rights reserved. Printed in the United States of America. No part of this publication may be reproduced, stored in a retrieval system, or transmitted, in any form or by any means, electronic, mechanical, photocopying, or otherwise, without prior written permission of the publisher, SAS Institute Inc.

Information in this document is subject to change without notice. The software described in this document is furnished under the license agreement printed on the envelope that contains the software diskettes. The software may be used or copied only in accordance with the terms of the agreement. It is against the law to copy the software on any medium except as specifically allowed in the license agreement.

First printing, February 1995

JMP®, JMP Serve®, SAS®, and SAS/INSIGHT® are registered trademarks of SAS Institute Inc. All trademarks above are registered trademarks or trademarks of SAS Institute Inc. in the USA and other countries. ® indicates USA registration.

Mac2Win® is a registered trademark of Altura Software, Inc.

Other brand and product names are registered trademarks or trademarks of their respective companies.

XTND Technology Copyright © 1989–1991 Claris Corporation. All rights reserved.

Portions of this software are Copyright © 1990 Aladdin Systems, Inc.

Mac2Win © software 1990-1994 Altura Software Inc. All rights reserved.

MacLinkPlus © 1984-1994 DataViz Inc. All rights reserved.

Portions of this software are copyrighted by Apple Computer Inc.

CONTENTS

Credits and Acknowledgments v

Chapter 1 1
    What You Need to Know 3
    Learning About JMP 3
    Manual Conventions and Organization 4
    Other Documentation 5

Chapter 2 7
    The File Menu 9
    The Edit Menu 18
    The Tables Menu 23
    The Rows Menu 36
    The Cols Menu 42
    The Analyze Menu 50
    The Graph Menu 56
    The Tools Menu 62
    The Window Menu 64
    The Help Menu (Microsoft Windows) 66

Chapter 3 69
    Elements of a JMP Data Table 71
    Creating a New Table 75
    Filling a Spreadsheet with Data 78
    Exporting JMP Files 87
    Characteristics of Data 88
    Printing and Journaling Data 93
    Using ClearAccess on the Macintosh 93

Chapter 4 95
    The Tables Menu 97
    The Group/Summary Command 98

The Subset Command	106
The Sort Command	107
The Stack Columns Command	108
The Split Columns Command	110
The Transpose Command	111
The Concatenate Command	114
The Join Command	117
The Table Info Command	125
The Attributes Command	125
The Design Experiment Command	129
Chapter 5	**131**
A Quick Example	133
The Calculator Window	135
Calculator Terminology	136
The Calculator Work Panel	138
The Formula Display	139
Keypad Functions	140
Function Browser Definitions	141
Chapter 6	**177**
Working with Formulas	179
Caution and Error Messages	185
Table Templates	188
Keyboard Shortcuts	191
Chapter 7	**193**
Report Windows	196
Graphical Displays	202
Help Windows	208
Copying from JMP Report Windows	213
Pasting Between Applications	215
Journaling Reports	215
Printing Reports and Journals	217
Appendix A Questions and Answers?	**219**
Appendix B What's New?	**223**
Index	**231**

Origin

JMP was 100% developed by SAS Institute Inc., Cary NC. The product consists of about 160,000 lines of C code. However, it is not a part of the SAS System and is not as portable as SAS. A SAS add-on product called SAS/INSIGHT is related to JMP in some ways, but has different conventions and capabilities. Portions of JMP were adapted from routines in the SAS System, particularly for linear algebra, and probability calculations. Version 1 of JMP went into production in October, 1989.

Credits

JMP was conceived and started by John Sall. Design and development was done by John Sall, Katherine Ng, Michael Hecht, Dave Tilley, and Richard Potter. Ann Lehman coordinated product development, production, and documentation. William Gjertsen, ambassador-at-large, interacts continuously with clients to provide current user feedback. Jeffrey Perkinson provides technical support and conducted test site administration. Annie Dudley headed testing for JMP Version 3, with contributions from Kristin Latour and the SAS Institute Quality Assurance department. Sales and sales support include Mary Ann Hansen, Kathryn Wise, and Chris Brown. Additional support is provided by Ruth Lee, Laura Wilson, Miranda Drake, Miriam Leyda, and Kelly Roeder, and past support by Kathy Kiraly and Russell Gardner.

The JMP manuals were written by Ann Lehman and John Sall with contributions and support from Kristin Latour, Mary Cole and Aaron Walker. Editing was done by Patricia Moell. Document production coordination was done by Curt Yeo. Production assistance included Patrice Cherry, Mike Pezzoni, Jennifer Albrecht and Lynn Friebel, graphic arts; Walt Martin, Postscript support; Aaron Walker, and Eric Gjertsen, indexing and help file implementation.

Special thanks to Jim Goodnight for supporting a product outside the usual traditions, and to Dave DeLong, for valuable ideas and advice on statistical and computational matters.

Thanks also to Robert N. Rodriguez and Ying So for statistical editorial support and statistical QC advice. Thanks to Georges Guirguis, Warren Sarle, Randall Tobias, Gordon Johnston, Ying So, Wolfgang Hartmann, Russell Wolfinger, Jane Pierce, and Warren Kuhfeld for statistical R&D support. Additional editorial support was provided by John Hansen, Marsha Russo, Dea Zullo, and Dee Stribling. Additional testing by Jeanne Martin, Fouad Younan, Jeff Schrilla, Jack Berry, Kari Richardson, Jim Borek, Kay Bydalek, Frank Lassiter. Additional technical support is provided by Jenny Kendall, Elizabeth Shaw, Mike Stockstill, and Duane Hayes. Thanks to Steve Shack, Greg Weier, and Maura Stokes for testing Version 1. Also thanks to Eddie Routten, John Boling, David Schlotzhauer, Donna Woodward, and William Fehlner.

Acknowledgments

We owe special gratitude to the people that encouraged us to start JMP, to the alpha and beta testers of JMP, and to the reviewers of the documentation. In particular we thank Al Best, Robert Muenchen, Stan Young, Lenore Hertzenberg, Morgan Wise, Frederick Dalleska, Stuart Janis, Larry Sue, Ramon Leon, Tom Lange, Homer Hegedus, Skip Weed, Michael Emptage, Mathew Lay, Tim Rey, Rubin Gabriel, Michael Friendly, Joe

Hockmen, Frank Shen, J.H. Goodman, Brian Ruff, and David Ikle. Also, we thank Charles Shipp, Harold Gugel, William Lisowski, David Morganstein, Tom Esposito, Susan West, Jim Winters, Chris Fehily, James Mulherin, Dan Chilko, Jim Shook, Bud Martin, George Fraction, Al Fulmer, Cary Tuckfield, Hal Queen, Linda Blazek, Ron Thisted, Ken Bodner, Donna Fulenwider, Nancy McDermott, Rick Blahunka, and Dana C. Aultman.

We also thank the following individuals for expert advice in their statistical specialties: R. Hocking and P. Spector for advice on effective hypotheses; Jason Hsu for advice on multiple comparisons methods (not all of which we were able to incorporate in JMP); Ralph O'Brien for advice on homogeneity of variance tests; Ralph O'Brien and S. Paul Wright for advice on statistical power; Keith Muller for advice in multivariate methods, Dave Trindade for advice on Weibull plots; Lijian Yang and J..S. Marron for bivariate smoothing design.

For sample data, thanks to Patrice Strahle for Pareto examples, the Texas air control board for the pollution data, and David Coleman for the pollen (eureka) data.

Technology License Notices

JMP software contains portions of the file translation library of MacLinkPlus, a product of DataViz Inc., 55 Corporate Drive, Trumbull, CT 06611, (203) 268-0030.

JMP for the Power Macintosh was compiled and built using the CodeWarrior C compiler from MetroWorks Inc.

SAS INSTITUTE INC.'S LICENSORS MAKE NO WARRANTIES, EXPRESS OR IMPLIED, INCLUDING WITHOUT LIMITATION THE IMPLIED WARRANTIES OF MERCHANTABILITY AND FITNESS FOR A PARTICULAR PURPOSE, REGARDING THE SOFTWARE. SAS INSTITUTE INC.'S LICENSORS DO NOT WARRANT, GUARANTEE OR MAKE ANY REPRESENTATIONS REGARDING THE USE OR THE RESULTS OF THE USE OF THE SOFTWARE IN TERMS OF ITS CORRECTNESS, ACCURACY, RELIABILITY, CURRENTNESS OR OTHERWISE. THE ENTIRE RISK AS TO THE RESULTS AND PERFORMANCE OF THE SOFTWARE IS ASSUMED BY YOU. THE EXCLUSION OF IMPLIED WARRANTIES IS NOT PERMITTED BY SOME STATES. THE ABOVE EXCLUSION MAY NOT APPLY TO YOU.

IN NO EVENT WILL SAS INSTITUTE INC.'S LICENSORS AND THEIR DIRECTORS, OFFICERS, EMPLOYEES OR AGENTS (COLLECTIVELY SAS INSTITUTE INC.'S LICENSOR) BE LIABLE TO YOU FOR ANY CONSEQUENTIAL, INCIDENTAL OR INDIRECT DAMAGES (INCLUDING DAMAGES FOR LOSS OF BUSINESS PROFITS, BUSINESS INTERRUPTION, LOSS OF BUSINESS INFORMATION, AND THE LIKE) ARISING OUT OF THE USE OR INABILITY TO USE THE SOFTWARE EVEN IF SAS INSTITUTE INC.'S LICENSOR'S HAS BEEN ADVISED OF THE POSSIBILITY OF SUCH DAMAGES. BECAUSE SOME STATES DO NOT ALLOW THE EXCLUSION OR LIMITATION OF LIABILITY FOR CONSEQUENTIAL OR INCIDENTAL DAMAGES, THE ABOVE LIMITATIONS MAY NOT APPLY TO YOU. SAS INSTITUTE INC.'S LICENSOR'S LIABILITY TO YOU FOR ACTUAL DAMAGES FOR ANY CAUSE WHATSOEVER, AND REGARDLESS OF THE FORM OF THE ACTION (WHETHER IN CONTRACT, TORT (INCLUDING NEGLIGENCE), PRODUCT LIABILITY OR OTHERWISE), WILL BE LIMITED TO $50.

Chapter 1
Preliminaries

JMP is statistical software that gives you an extraordinary graphical interface to display and analyze data. JMP is for interactive statistical graphics and includes

- a spreadsheet for viewing, editing, entering, and manipulating data
- a broad range of graphical and statistical methods for data analysis
- options to highlight and display subsets of the data
- data management tools for sorting and combining tables
- a calculator for each table column to compute values
- a facility for grouping data and computing summary statistics
- special plots, charts, and communication capability for quality improvement techniques
- tools for moving analysis results between JMP and other applications and for printing
- a scripting language for saving frequently used routines.

JMP is easy to learn. Statistics are organized into logical areas with appropriate graphs and tables, which help you find patterns in data, identify outlying points, or fit models. Appropriate analyses are defined and performed for you, based on the types of variables you have and the roles they play.

The graphs and reports are dynamically linked. For example, when you click a point in a plot, the point label appears, and its corresponding row highlights in the spreadsheet window and in all other plots where that point is represented.

JMP offers descriptive statistics and simple analyses for beginning statisticians and complex model fitting for advanced researchers. Standard statistical analysis and specialty platforms for design of experiments, statistical quality control, ternary and contour plotting, and survival analysis provide the tools you need to analyze data and see results quickly.

Chapter 1
Contents

What You Need to Know 3
Learning About JMP..3
Manual Conventions and Organization 4
Other Documentation 5

What You Need to Know

...about your computer
Before you begin using JMP, you should be familiar with standard operations and terminology such as *click, double-click, COMMAND-click* and *OPTION-click* on the Macintosh (*CONTROL-click* and *ALT-click* under Windows), *SHIFT-click, drag, select, copy,* and *paste.* You should also know how to use menu bars and scroll bars, move and resize windows, and manipulate files. If you are using your computer for the first time, consult the reference guides that came with it for more information.

...about Statistics
Even though JMP has many advanced features, you need only a minimal background of formal statistical training. All analysis platforms include graphical displays with options that help you review and interpret the results. Each platform also includes access to help windows that offer general help and some statistical details.

Learning About JMP

...on your own with JMP Help
If you are familiar with Macintosh or Microsoft Windows software, you may want to proceed on your own. After you install JMP, you can open any of the JMP sample data files and experiment with analysis tools. Help is available for most menus, options, and reports.

There are several ways to see JMP Help:

- Select **About JMP** from the Apple menu on the Macintosh or the **Contents** command from **Help** menu of Microsoft Windows. Your cursor becomes a question mark as it passes over the list of buttons to the right of the About JMP window. When you click on a button, JMP leads you through help about that topic.
- Click the **Statistical Guide** button on the Macintosh About JMP screen, or select **Statistical Guide** from the **Help** menu under Windows. The JMP Statistical Guide is a scrolling alphabetical reference that tells you how to generate specific analyses using JMP and accesses further help for that topic.
- You can choose **Help** from pop-up menus in JMP report windows.
- After you generate a report, select the help tool (**?**) from the **Tools** menu and click the report surface. Context-sensitive help tells about the items in that report window.
- If you are using Microsoft Windows, help in typical Windows format is available under the **Help** menu on the main menu bar.
- If you use a Macintosh with Version 7 system software, balloon help is available.

...hands–on tutorials

The *JMP Introductory Guide* is a collection of tutorials designed to help you learn JMP strategies. Each tutorial uses a file from the Macintosh SAMPLE DATA folder, or the file called data in the JMP directory under Windows. By following along with these step–by–step examples, you can quickly become familiar with JMP menus, options, and report windows.

...reading about JMP

The book you are reading now is the *JMP User's Guide*. It gives you reference material for all JMP menus, an explanation of data table manipulation, a description of the calculator and how to use it, and a discussion of scripting to capture and run a sequence of actions. See the *JMP Statistics and Graphics Guide* for documentation of the **Analyze** and **Graph** menus, and for the experimental design facility in JMP.

...if you are a previous JMP user

If you are familiar with the JMP environment, you may want to know only what's new. Appendix B gives a summary of general changes and additions in Version 3.1 of JMP.

Manual Conventions and Organization

The following manual conventions help you relate written material to information you see on your screen:

- Most open data table names used in examples are capitalized (ANIMALS or ANIMALS.JMP) in this document. In the Macintosh SAMPLE DATA folder and the file names it conains are capitalized (ANIMALS). Files in the Windows data folder are in lower case (animals.jmp).
- Note→ Special information, warnings, and limitations are noted as such and followed with an arrow as shown in this note.
- Reference to menu names (**File** menu) or menu items (**Save** command) appear in the **Helvetica bold** font, similar to the way they appear on your screen.
- References to variable names in data tables and items in reports show in either Helvetica or **Helvetica bold**, according to the way they appear on the screen or in the book illustration.
- Words or phrases that are important or have definitions specific to JMP are in *italics* the first time you see them.

The chapters and appendixes in this book are organized as follows:

Chapter 2, The Menu Bar

    documents the main menu bar and describes each menu command.

Chapter 3, Using JMP Data Tables
: explains how to create new JMP tables; import and export data; navigate the table to modify values; add and delete rows and columns; cut, paste, and paste at end; print; and save JMP table information.

Chapter 4, The Tables Menu
: explains the **Tables** menu commands with examples of how to use the **Group/Summary** menu command; and shows how to sort, subset, transpose, join, and concatenate JMP tables.

Chapter 5, Calculator Functions
: introduces the JMP calculator, describes the calculator components, and lists all the functions with examples.

Chapter 6, Using the Calculator
: explains in detail how to use the JMP calculator, create formulas, modify formulas, use template tables, and how to create your own table templates.

Chapter 7, Report Windows
: describes the form and features of JMP results in report windows.

Appendix A
: is a collection of common questions and answers.

Appendix B
: gives an overview of new features in Version 3 of JMP.

Other Documentation

The *JMP Introductory Guide* has beginning and advanced tutorials.

The *JMP Statistics and Graphics Guide* documents the **Analyze** menu commands, the **Graph** menu commands, and the experimental design facility in JMP.

On the Macintosh, JMP can import real–time data using the Macintosh Communications Toolbox, which comes with supplemental documentation. If you do not already have this communications software or if you need the supplemental documentation contact, the JMP Sales Department at SAS Institute Inc.

Chapter 2
The Menu Bar

When you open the JMP application, the first thing you see is the menu bar and an empty untitled data table as shown in **Figure 2.1**.

Figure 2.1 The Menu Bar and Empty Data Table

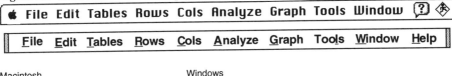

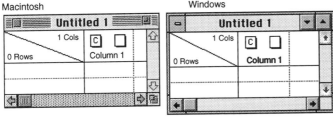

This chapter contains documentation for all the commands in the JMP menu bar. It is organized for easy reference. The menus and commands are presented in their order on the menu bar. Each menu title shows beneath the page number in its page header to help you locate specific commands quickly.

Chapters 3 through 7 give information and examples about data table management, a description of the calculator with examples of how to use it, a discussion and examples of the **Tables** menu commands, and details about report windows.

If you prefer to begin using JMP with a hands–on approach, you can skip this chapter and begin with Chapter 3, "JMP Data Tables," or refer to the *JMP Introductory Guide* for tutorials.

Chapter 2
Contents

The File Menu 9
The Edit Menu 18
The Tables Menu 23
The Rows Menu 36
The Cols Menu 42
The Analyze Menu 50
The Graph Menu 56
The Tools Menu 62
The Window Menu 64
The Help Menu (Microsoft Windows) 66

The File Menu

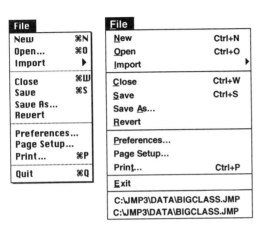

The **File** menu has commands that perform file management or affect the JMP environment. Related commands are grouped according to their functions. The four groups contain commands that

- open data files and create new data tables
- close windows, close and save data tables, and retrieve a saved data table
- set global options and print JMP windows
- end a JMP session.

You modify a data table by working with its spreadsheet view as the active window. In descriptions of menu commands *data table*, *table*, and *spreadsheet* are used synonymously unless specifically noted otherwise.

The **New** command creates and displays a new empty untitled data table with one column labeled Column 1. To create more columns use **Rows** and **Cols** menu commands and type or paste data into the spreadsheet. Use **Save As** from the **File** menu to name the table and save it as a JMP file. Use the **Open** command to reopen the file after it has been saved.

The **Open** command displays the standard open file dialogs. You can open an existing JMP file using this command, or you can open a file by double clicking it. When you open a file, you have access to the spreadsheet view of the data table.

Figure 2.2 Standard Open File Dialog

Macintosh

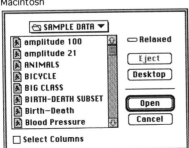

Windows

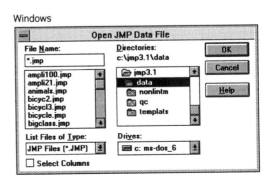

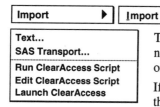

The **Import** command lets you import raw data and create a new JMP data table by reading data from a text–format file or a SAS transport file.

If you have the ClearAccess query tool on your Macintosh, three additional **ClearAccess** commands also show.

The **Import** command pop–up selections display modified open file dialogs appropriate for the selection you choose:

The **Text** Option

displays the dialog shown in **Figure 2.3**. It creates a new JMP data table by reading data from a text–format file. When you import a file in text format, you can use options to identify the field and line delimiters contained in the incoming file.

By default, the Text option in the **Import** command reads a text file of rows and columns without header information and names the JMP data table columns Column 1, Column 2, and so on.

If the file has labels, JMP header information, special characters to denote end of fields or end of lines, or if data values are enclosed in quotes, use the radio buttons and check boxes in the Import dialog described next:

⦿ **Labels**

tells JMP the incoming data are from a text–format file that contains column labels as the first record of the data. These labels become column names in the new JMP data table.

⦿ **JMP Export**

indicates that the incoming data are from a text format file that contains column header information as exported from a JMP file.

The check boxes in the Import dialog identify the end–of–field and end–of–line delimiters.

End of Field

designates the character that identifies the end of each field in the incoming file. This delimiter can be a tab, one or more blank spaces, any character, or any combination of the above. When you choose **Other**, an editable text box appears where you can enter a character or its hexadecimal representation.

End of Line

selects the characters or keys that identifies the end of each line. You can use the return key, the line feed character, a semicolon, or another character of your choice. The **Other** option works the same as it does with **End of Field** described above.

Note➜ To enter a hexadecimal value, precede the value with '0x'.

Strip Enclosing Quotes

tells JMP that each data value is enclosed in single or double quotes. JMP removes the quotes before importing the data.

Figure 2.3 The Text Import Dialog

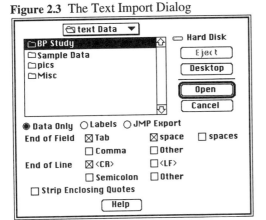

Macintosh Text Import Dialog

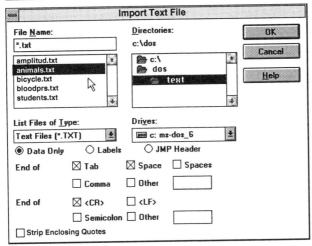

Windows Text Import Dialog

The **SAS Transport** Option

displays the dialogs in **Figure 2.4**. A *SAS transport file* is a portable file created by the SAS System. If you click the **Open All Members** box, JMP creates a data table for each member in the SAS data library. Appendix A, "Questions and Answers," tells you more about how to create a SAS transport file and import it into JMP.

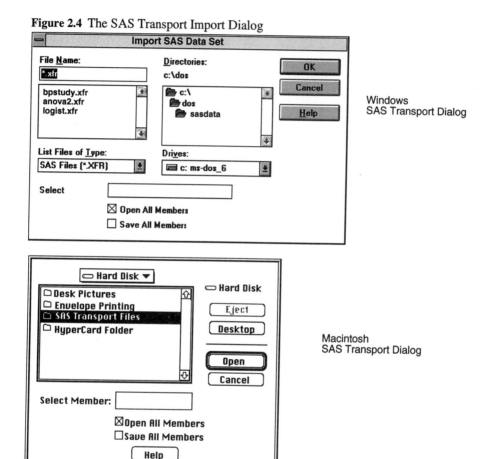

Figure 2.4 The SAS Transport Import Dialog

The ClearAccess Query Tool for the Macintosh

If you have the ClearAccess query tool from ClearAccess Corporation installed on your Macintosh, the **Import** command also lists **ClearAccess** commands. JMP Version 3.1 for the Macintosh includes special support for ClearAccess. The ClearAccess package allows you to create, edit, and submit database queries to a variety of popular industry–standard relational databases. The ClearAccess commands give you direct importing of data from a remote database. JMP executes ClearAccess scripts and the resulting data appears in a new JMP data table with no intermediate steps.

The ClearAccess commands are briefly described here. For details about writing, editing, and executing ClearAccess scripts, see the documentation shipped with ClearAccess.

Run ClearAccess Script

prompts the user for the name of an existing ClearAccess script and asks the ClearAccess application to execute it. If the ClearAccess application is not running, it is launched and brought to the foreground for the duration of the query. During the query, you will occasionally see status messages displayed in the ClearAccess application's window. When the script finishes execution, any data extracted from the remote database is placed into a new JMP data table named **ClearAccess Import**. You can rename the table, and use it as you would any other JMP table.

Edit ClearAccess Script

brings the ClearAccess application to the foreground and displays an editing window. You can then edit your script and save the changes. Details on writing ClearAccess scripts can be found in the ClearAccess documentation.

Launch ClearAccess

is a shortcut for users who need to use the ClearAccess application, but do not want to execute or edit a script immediately. When choosing this option, the ClearAccess application is launched and is brought to the foreground.

The **Close** command closes the active window. When you close a spreadsheet window while any of its analysis windows are open, a dialog asks if you want to close the data table only or all related analysis windows as well. If you close the spreadsheet, a dialog asks you whether to save or discard changes to the table. The **Close** command is the same as clicking the close box of the active window.

The **Save** command writes the active data table to a file. If the data table has been saved before, it is rewritten to the same file name and location, replacing the old information. If the data table is new, the **Save** command has the same effect as the **Save As** command. Saving a data table does not automatically close it.

Note→ JMP report windows are not saved with the data table. However, you can use JMP tools to copy reports to other applications, or you can use the journaling feature to save reports in text or word processing format.

The **Save As** command writes the active data table to a file after prompting you for a name and disk location, as shown in **Figure 2.5**. It saves the data table as a JMP file, or converts it to another format if you click a different radio button in the dialog.

14 Chapter 2 *The Menu Bar*
File

Figure 2.5 Save As Dialog and Text Formatting Options

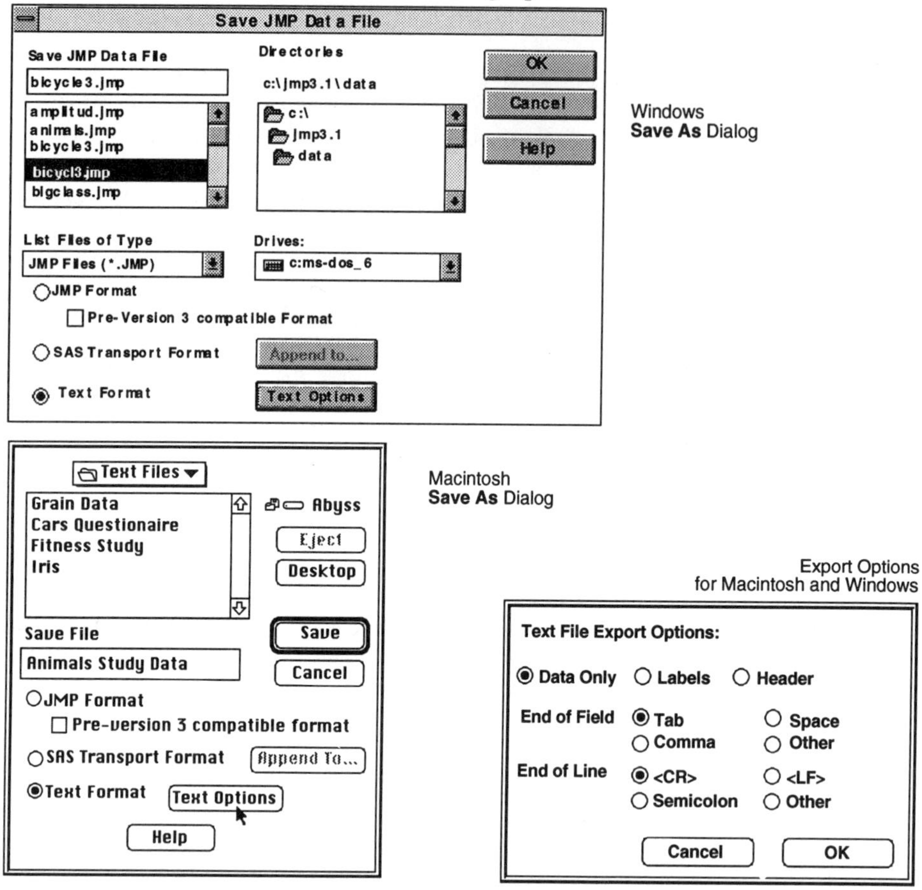

Use the options and radio buttons for exporting JMP files as needed:

- ⦿ **JMP format**
  saves the JMP data table in JMP format.

- ☐ **Pre-Version 3 compatible format**
  saves the JMP table in a form that previous releases of JMP can read. If this option is not checked, the saved table is accessible only by JMP Version 3.

- ⦿ **SAS Transport**
  converts a JMP data table to SAS transport file format and saves it in a SAS transport library.

  Note→ If you don't use the **Append To** option, a new SAS transport library is created using the name and location you give. If you do not specify a new file name, the SAS transport library replaces the existing JMP data table.

Append To...
appends the active data table to an existing SAS transport library.

⦿ **Text Format**
converts data from a JMP file to standard text format, maintaining the rows and columns. For both the Macintosh and Windows the **Text Format** option displays the dialog shown at the lower right in **Figure 2.5**, which has choices to describe specific text arrangements:

⦿ **Data Only** requests that no labels or header information be saved with the data.

⦿ **Labels** specifies that JMP column names be written as the first record of the text file.

⦿ **Header** exports JMP header information and data to a text file. The header information includes the number of rows and columns in the file and information about each column. If you need the data in JMP format again, JMP imports this kind of text file more efficiently than a standard text file, using the **JMP Format** radio button in the Import dialog.

End of Field and **End of Line**
designate the characters to identify the end of each field and end of line in the saved text file. These options are described previously in the **Import** command.

| Revert | Revert |

The **Revert** command restores the data table to the form it had when last saved. If it was last saved as a SAS Transport file or a text format file, the **Revert** command is not available. In this case, use the **Import** command to restore the data table.

| Preferences... | Preferences... |

The **Preferences** command displays the dialog shown in **Figure 2.6** to set and save a tailored JMP environment.

Figure 2.6 The Preferences Dialog

Macintosh Preferences Dialog

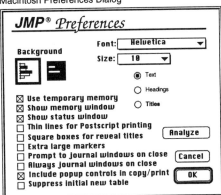

Windows Preferences Dialog

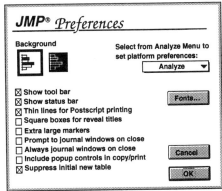

You can check any combination or all of the following preferences:

- The background for all JMP graphical displays except the Spinning Plot platform can be either white or black. You change them with the **Background** icons. The Spinning platform has an independent background setting.

- On the Macintosh the **Font** and **Size** pop-up menus let you specify any font, font style, and font size available in your system for the text, headings, and titles of JMP tables and report windows. Under Windows, the Font button displays the dialog in **Figure 2.7** to select font, font style, and font size. In addition, the sample text shows what these selections look like.

Figure 2.7 The Font Specification Dialog Under Windows

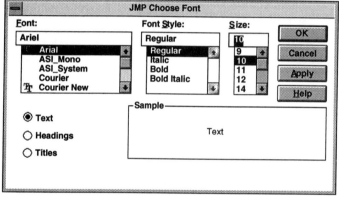

- On the Macintosh, when you click the first three check boxes, memory information shows in the lower-left corner of your desktop.

- Under Windows, when you click the first check box, the tool bar shown below displays beneath the main menu bar. It has icons for the most commonly used **File** menu commands, and icons for all **Analyze**, **Graph**, and **Tools** menu commands

- Under Windows, when you click the second check box, memory information, a brief sentence of help, and the current time show in a status bar across the bottom of your JMP window.

- **Thin Line for Postscript printing** reduces the lines used in JMP reports from 1 point to .5 points. The line thickness reduction does not show on your screen, but appears when you print JMP results.

- **Square box for reveal titles** changes reveal/conceal buttons and the boundary frames of graphical displays and text reports from round rectangles to rectangles.

- **Extra large markers** increases the size of all markers in all plots.

- **Prompt to journal window on close** displays a dialog that prompts you to journal your report whenever you close an analysis report window.
- **Always journal window on close** automatically journals a report window when you close it. The **Journal** command is discussed in **The Edit Menu** section later in this chapter.
- **Include popup controls in copy/print** copies to the clipboard or prints pop–up menu icons with reports. By default, the **Copy** and **Print** commands do not show these icons.

Note→ The Preferences **Analyze** button displays the **Analyze** menu commands in a pop–up menu. When you select an **Analyze** command, a dialog lists that platform's text reports and graphical displays. To tailor the initial display of an analysis platform, select the options you want, and then click **Save**. Whenever you select an **Analyze** command, the window components you saved as preferences automatically display.

The **Page Setup** command displays the standard dialog for setting printed page characteristics. The form of the dialog depends on your current printer driver.

The **Print** command prints the active window. It displays the standard dialog for printing. The appearance of this dialog depends on your printer driver.

On the Macintosh, the **Quit** command closes all JMP windows (prompting you to save changes) and quits the JMP application.

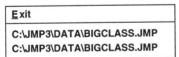

Under Windows, the **Exit** command closes all JMP windows (prompting you to save changes) and quits the JMP application.

Also, following the Exit command, the **File** menu lists the four JMP tables most recently opened. When you click on a table's name in this list, JMP re–opens the table if it is closed or gives you a new spreadsheet view if it is currently open.

The Edit Menu

The **Edit** menu contains standard commands including **Cut** and **Copy**. These commands operate on rows and columns that are selected as shown in **Figure 2.8** on selected areas of reports, and on selected formula elements in the calculator.

In the data table spreadsheet, you can select rows, columns, or both rows and columns at the same time. **Edit** commands operate on entire rows if no columns are selected. Likewise, they operate on whole columns if no rows are selected. If you select both rows and columns, **Edit** commands affect the subset of values defined by the intersection of those rows and columns.

To select a row in a spreadsheet, click the space that contains the row number. To select a column, click the white background area. To select multiple rows or columns, drag across them or SHIFT–click the first and last rows or columns of a range. To make a discontiguous selection use COMMAND–click on the Macintosh or CONTROL–click under Windows to highlight the rows and columns you want. To select a block of cells formed by the intersection of rows and columns, drag the cross cursor diagonally across the subset of cells.

Note→ You can also select special subsets of rows using the **Select** command in the **Rows** menu, described later in this chapter.

Figure 2.8 Selecting Rows and Columns

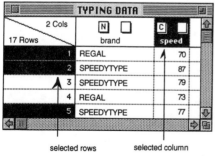
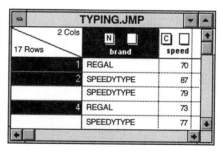

You can also use the **Edit** menu in conjunction with JMP tools to copy all or part of active analysis report windows. See the **Tools** menu section for more information.

| Undo   ⌘Z | | Undo   Ctrl+Z |

The **Undo** command cancels the effect of the most recent reversible **Edit**, **Rows**, or **Cols** command. If **Undo** is available, its selection in the **Edit** menu appears as **Undo** *command* where *command* is the most recent action. Most destructive spreadsheet operations (such as cut, paste, or delete rows) are reversible. **Undo** dims when the most recent command is irreversible. Once you select **Undo**, its menu item changes to **Redo** *command*, where *command* is the action that was undone. **Redo** cancels the effect of **Undo**.

| Cut   ⌘X | | Cut   Ctrl+X |

The **Cut** command copies selected fields from the active spreadsheet to the clipboard and replaces them with missing values. It is equivalent to **Copy**, then **Clear**. You can also use the **Cut** command to copy all or part of a report window. However, **Cut** works like **Copy** in graphical displays (it does not clear the copied image).

| Copy   ⌘C | | Copy   Ctrl+C |

The **Copy** command copies the values of selected data cells from the active data table to the clipboard. If you do not select columns, **Copy** copies entire rows. Likewise, you can copy values from whole columns if no rows are selected. If you select both rows and columns, **Copy** copies the subset of cells defined by their intersection.

You can also use the **Copy** command to capture graphical displays or text reports defined by the scissors tool in the **Tools** menu.

Note→ Data you cut or copy to the clipboard can be pasted into JMP tables or into other applications. Pictures can be pasted into any application that accepts graphics.

| Copy as Text | | Copy as Text |

The **Copy as Text** command copies all text from the active report window (no graphical displays) and stores the text in the clipboard.

If you select **Copy as Text** while holding down the option key (OPTION–**Copy as Text** on the Macintosh or ALT–**Copy as Text** under Windows), the column names in the JMP table are written to the clipboard as the first line of information. If you use OPTION or ALT with **Copy as Text** to copy a text report, column names in the report are written to the clipboard as the first line of information.

| Paste   ⌘V | | Paste   Ctrl+V |

The **Paste** command copies data from the clipboard into a JMP data table. **Paste** can be used with the **Copy** command to duplicate rows, columns, or any subset of cells defined by selected rows and columns.

Edit

To duplicate an entire row or column,

- Select and **Copy** the row or column to be duplicated.
- Select an existing row or column to receive the values.
- Use the **Paste** command.

To duplicate a subset of values defined by selecting both rows and columns. Follow the steps above but select the same arrangement of rows and columns to receive the copied values as originally contained them. If you paste data with fewer rows into a destination with more rows, the source values recycle until all receiving rows are filled.

| Paste at End | Paste at End |

The **Paste at End** command extends a JMP data table by adding rows and columns to a data table as needed to accept values from the clipboard.

If you select rows before choosing this command, the effect is similar to joining data tables by row number. If you highlight columns, the **Paste at End** command adds cells to the bottom of the data table filling them with values from the clipboard.

To transfer data from another application into a JMP data table, first copy the data to the clipboard from within the application. Then use the **Paste at End** command to copy the values to a JMP data table. Rows and columns are automatically created as needed when you **Paste at End**.

If you select **Paste at End** while holding down the OPTION or ALT key with **Paste at End**, the first line of information on the clipboard is used as the column names in the new JMP data table.

See Chapter 3, "JMP Data Tables," for more information about **Paste at End**.

| Clear | Clear |

The **Clear** command clears all selected cell values from the active data table and replaces them with missing values. The values are not copied to the clipboard.

| Journal ⌘J | Journal Ctrl+J |

The **Journal** command copies all information from the active data table or report window to an open journal window. Each subsequent **Journal** command appends a page break followed by the information from the active JMP window to the current journal file. You can journal as much information as you need into a single open journal window. The scissors tool in the **Tools** menu and the **Cut** and **Paste** commands in the **Edit** menu let you cut and paste sections of JMP analysis reports into the journal window. You can also type your own notes to the end of the open journal file at any time.

When you select the **Save As** command for an active journal window, you give a name and disk location for that journal file. The Save As dialog has a pop–up menu that lists the available word–processing formats for saving journal information.

Chapter 2 *The Menu Bar* **21**

Edit

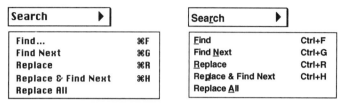

The **Search** command gives you the ability to find and replace text in the usual way found in most word processing and editing programs. Find and replace actions deal only with character strings. Numbers are treated as text, and appear to the **Search** command as they show in the data table. All dialog actions also respond to the keystroke shortcuts shown on the **Search** pop–up menu. For multiple Search actions, it is sometimes much more convenient to use the keystroke shortcuts instead of the **Search** dialog.

If there is no value in the **Search** dialog **Find What** box, all **Search** pop–up menu items except **Don't Find, Cancel** and **Help** are inactive (dimmed).

The **Search** pop–up menu selections function as follows:

Find

displays the dialog shown in **Figure 2.9**, which prompts you to enter a *find value* in the **Find What** box. Optionally, you can enter a replace value in the **Replace With** box. After you enter a *find value* and click **Find**, JMP searches the active table for the *find value*.

The search begins in the focused cell. A cell is focused when it is highlighted or contains the blinking vertical bar that indicates an insertion point. By default (if there is no focused cell), the search begins with the first cell in the first column. The search covers every table cell until it locates a *find value* or reaches the end of the table. The *find value* highlights when located and you hear a beep when you reach the end of the table, or return to the cell where you began your search.

The **Find** command does not find values in locked or hidden columns, or locked tables. To find these values, you must either unhide the column, or unlock the column or table. Values in columns with formulas can only be found by using the **Rows: Select Where** command.

Figure 2.9 Find and Replace Dialog

Find Next
: continues to search for the *find value* by selecting **Find Next** in the **Search** pop–up menu again, by clicking **Find** in the dialog, or by using COMMAND–G on the Macintosh or CONTROL–G under Windows.

Replace
: replaces the currently highlighted cell value with the contents of the **Replace With** box (the *replace value*) in the **Find** dialog. If the *replace value* is missing and you select Replace or use COMMAND–R on the Macintosh or CONTROL–R under Windows, the currently highlighted cell content becomes a missing value.

Replace & Find Next
: functions the same as **Replace**, but continues to search for the find value.

Replace All
: replaces all occurrences of the *find value* with the *replace value*.

Match Case
: The **Match Case** check box gives you a case sensitive search, useful for locating proper nouns or other capitalized words.

Check boxes on the Find and Replace dialog give additional options:

Match whole words only
: Blanks count as characters, which lets you search for a series of words in a character column, or locate strings with unwanted leading or trailing blanks. Using the **Match whole words only** check box also locates words with at least one leading and one trailing blank.

Search by row
Search by column
: **Search by row** searches the data table row by row from left to right, to the rightmost cell in the last row or until you stop the search. Likewise, **Search by column** searches the table column by column, from top to bottom until it reaches the last cell in the rightmost column, or until you stop the search. If you begin a search with a focused cell, as describe above, you limit the search to the following rows or columns according to the respective radio button in effect.

Note➡ the **Undo** command works with **Replace**, **Replace & Find Next**, and **Replace All**.

The Tables Menu

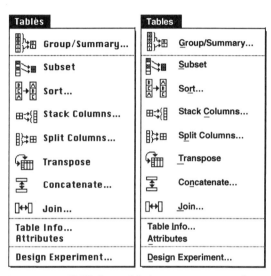

The **Tables** menu commands modify or create a new JMP table from one or more existing tables:

Group/Summary creates a JMP window that contains a *summary table*, which summarizes columns from the active data table, called its *source table*. The summary table has a single row for each level of a grouping variable you specify. Optionally, you can add columns of summary statistics to this table at any time.

Subset creates a new data table formed by the selected rows and columns in the active spreadsheet.

Sort sorts a JMP data table by one or more columns.

Stack Columns creates a new data table by stacking specified columns from the active data table into a single column in the new data table.

Split Columns creates a new data table by dividing one or more specified columns into several new columns according the values of one or more ID variables.

Transpose creates a new JMP table that has the columns of the active table as its rows and the rows of the active table as its columns.

Concatenate creates a new data table from two or more open data tables by combining them end to end.

Join creates a new data table by merging (joining) two tables side by side.

Table Info displays a dialog for you to make notes about the current data table or to examine its attributes.

Attributes creates a special data table that contains editable information about the current data table. The *attributes table* has a row for each variable in its source table and variables for all column characteristics. Editing characteristics in the attributes table changes them in the source table when you select the attributes table **Update** command.

Design Experiment accesses the design–of–experiments module of JMP. Chapters 25 through 30 of the *JMP Statistics and Graphics Guide* document design of experiments.

This section gives you an overview of each **Tables** menu command. Dialogs used by the commands are also described here. See Chapter 4, "The Tables Menu," for details and examples of each **Tables** command.

24 Chapter 2 *The Menu Bar*

Tables

The **Group/Summary** command creates a JMP window that contains a *summary table*. This table summarizes columns from the active data table, called its *source table*. The summary table has a single row for each level of a grouping variable you specify. When there are several grouping variables, the summary table has a row for each combination of levels of all grouping variables. The summary table is linked to its source table. When you highlight rows in a summary table, the corresponding source table rows highlight.

Figure 2.10 A Source Table with a Summary Table

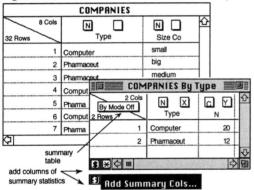

The illustration in **Figure 2.10** shows a source table called COMPANIES (or COMPANYS.JMP under Windows), with a summary table grouped by type of company. The Type column in the source table has the values Computer and Pharmaceut. The COMPANIES By Type summary table has two rows that show counts of 20 computer companies and 12 pharmaceutical companies.

Initially, the summary table lists only the grouping values (levels) and their frequency in the source table. However, you can use the summary table for the following purposes:

Create a table of summary statistics

The **Stats** pop–up menu on the Group/Summary dialog lists standard univariate descriptive statistics. With this menu you can add columns of descriptive statistics to the summary table for any numeric column in the source table. You can also access the Group/Summary dialog from an existing summary table with the **Add Summary Cols** command in the dollar ($) pop–up command (see **Figure 2.10**).

Analyze subsets of data

An active summary table has the two *modes*, accessed by the pop–up menu at the upper left corner of the table:

By Mode Off — When By Mode is off, the **Analyze** and **Graph** menu commands, and the **Transpose** command in the **Tables** menu apply to the active summary table itself instead of its source table.

By Mode On — When By Mode is on, highlighted rows in the active summary table identify subsets in the source table. Commands on the **Analyze** and **Graph** menus, and the **Transpose** command in the **Tables** menu recognize these subsets. A single **Analyze** or **Graph** menu command produces a separate report window for each subset identified by a highlighted row in this summary table.

Plot and chart data
> The **Bar/Pie Charts** command in the **Graph** menu displays summarized data. The Bar/Pie Charts platform requires that an X variable or combination of X variables identify unique groups. You may need to process data tables with the **Group/Summary** command and produce summary tables to use this platform. The Bar/Pie Charts platform is discussed later in this chapter and in Chapter 18, "Bar, Pie, and Line Charts," of the *JMP Statistics and Graphics Guide*.

When you select **Group/Summary**, you see the dialog in **Figure 2.11**.

Figure 2.11 The Group/Summary Dialog

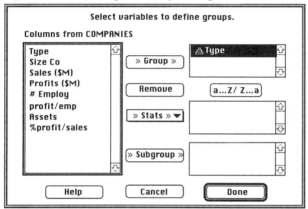

To complete the Group/Summary dialog select variables and click the appropriate button. On the Macintosh you can drag the selection to the list box you want to use. To create a summary table, use components and options in the Group/Summary dialog:

Columns from *tablename*
> To specify grouping variables, select one or more variables from the **Columns** list and click **Group**. The summary table that results has one row for each combination of levels of all specified grouping variables. The order in which you build the grouping list establishes the order of the rows in the summary table. The rows in the summary table list the grouping variable level combinations in standard sorting sequence.

Remove
> The **Remove** button removes highlighted columns from any list on the right side of the Group/Summary dialog.

a...Z/Z...a
> You can specify that the values of any column summarize in either ascending or descending order by selecting that column in the grouping variable list and clicking on the **a...Z/Z...a** toggle. By default, column values are grouped in ascending order.

Stats
> The **Stats** pop–up menu, shown in **Figure 2.12**, can add columns of statistics to a summary table. To add summary columns, highlight any numeric column in the column selection list and select the statistic you want from the **Stats** pop–up menu. When you release the mouse, the selection shows in the middle column list of the

Group/Summary dialog. A new column of statistics with that name appears in the summary table.

Figure 2.12 Adding Summary Statistics

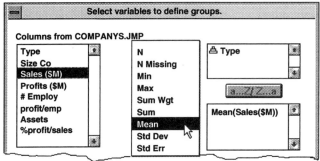

Subgroup
If you choose a subgroup variable, a column of statistics is added to the summary table for each level of the subgroup variable along with overall statistics for levels of the grouping variables.

The summary table, like like analysis results, is not saved when you close it. However, the **Save As** command saves the summary table as a standard JMP document. Or, if you select the **Subset** command when the summary table is the active window, the result is a standard untitled JMP data table. This new untitled table is a duplicate of the summary table, but it is not linked to the source table.

See Chapter 4, "The Tables Menu," for examples of how to group data and create columns of summary statistics. Chapter 19, "Overlay Plots," in the *JMP Statistics and Graphics Guide* shows examples of graphing summary information.

The **Subset** command produces a new JMP table that is a subset of the active data table. The new table has the rows and columns defined by the highlighted rows and columns from the active spreadsheet. Highlighting rows is done by
- selecting them on the spreadsheet
- highlighting histogram bars or points on a plot generated by the data table
- using one of the **Select** commands from the **Rows** menu
- selecting one or more rows in a summary table produced by the **Group/Summary** command (described previously) to define a subset in its source table.

Note➡ When you hold down the SHIFT key and select **Subset**, the subset table that results and any plot or graph of that subset table remain linked to the original table. Highlighting rows in this kind of subset table highlights the corresponding rows in the original table and in all its plots and graphs.

The **Sort** command sorts a JMP data table by one or more columns. The **Sort** command displays the dialog shown in **Figure 2.13**, for you to specify columns as sort fields. Click **Sort** after you complete the dialog.

Figure 2.13 Sort Field Selection Dialog

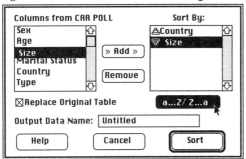

Use components and options in the **Sort** dialog as follows:

Columns from *table name*
Select sort fields from the **Columns** list and add them to **Sort By** list with the **Add** button. If you are using a Macintosh you can drag variables from one list to another. The order in which you build this list establishes the sorting order.

Sort By
The **Sort By** list shows the columns to be sorted in order of precedence. You can remove sort fields by selecting them in the sort list and clicking the **Remove** button.

a...Z/Z...a
You can sort a column in either ascending or descending order by it in the Sort By list and clicking the **a...Z/Z...a** toggle. By default, columns sort in ascending order.

Replace Original Table
By default, **Sort** creates a new data table. You can replace the original data table by checking **Replace Original Table** as shown in **Figure 2.13**.

Note➡ Formulas are not preserved in the new sorted table.

The **Stack Columns** command creates a new data table from the active table by stacking specified columns into a single new column. The values in other columns are preserved in the new data table. In addition, **Stack** creates an ID column to identify each row in the new table.

The Stack dialog shown in **Figure 2.14** specifies that two columns from the POPCRNTR.JMP table (POPCORN TRIALS on the Macintosh), called **trial 1** and **trial 2**, *stack* into a single column assigned the name **yield**. The values of the ID column are the names of the columns in the original table that are to be stacked (trial 1 and trial 2).

Figure 2.14 Stack Columns Dialog

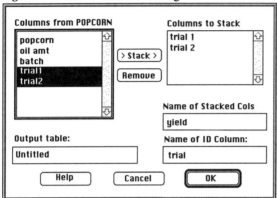

You use the elements of the Stack dialog as follows:

Columns from *table name*
Select columns to be stacked from this list and add them to the **Columns to Stack** list by clicking or by dragging them to the list box (Macintosh only).

Columns to Stack
All columns in the **Stacking Columns** list stack or pile one upon another in the order they appear in this list. You can remove a column from the list by selecting it and clicking **Remove**. The number of rows in the new table is the total number of stacked columns multiplied by the number of rows.

Name of Stacked Cols
Enter a name in the **Stacked Column Name** box to name the new column of stacked values. By default, this column name is _Stacked_. If you delete the default name and do not specify a new name, the new stacked column is not written to the data table.

Name of ID Column
Enter a name in the **Type Column Name** box to name the column whose values identify the original column for each stacked cell in the new table. By default, this column name is _ID_. If you delete the default name and do not specify a new name, the new ID column is not written to the data table.

Output Table Name:
Optionally, you can specify a table name to use instead of the default name, **Untitled**.

The **Split Columns** command creates a new data table from the active table by splitting one or more columns to form multiple columns. The new columns correspond to the values (levels) of an ID variable. **Split Columns** displays the dialog shown in **Figure 2.15**. for you to specify the columns to be split and the ID variable. **Stack** also requires one or more columns whose combined values identify each row in the new table. Optionally, the values in the other columns can be preserved in the new data table.

Figure 2.15 The Split Columns Dialog

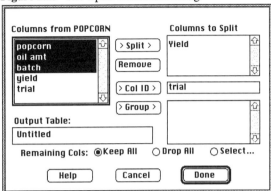

The Split Columns dialog example in **Figure 2.15** specifies that the yield column values in the POPCORN (POPCORN.JMP) data table split to form a column for each level of trial, the ID variable. The values of trial (trial1 and trial 2) become the names of the new columns. You use the Split Column dialog as follows:

Columns from *table name*
  lists the columns in the current data table.

Split button and the **Columns to Split** list box
  Select one or more columns that you want to split (*unstack*) to form multiple columns according to the values (levels) of selected grouping variables, or to the levels of the **Col ID** variable. Click **Split**, or drag the selection to the **Columns to Split** list box.

Col ID
  A column ID variable is required. Select a single column from the active data table whose values become names for the new split columns. If there are no **Group** variables specified, the levels of the **Col ID** variable also define the split levels. Click **Col ID**, or drag the selection to the **Col ID** box.

Group
  Optionally, select one or more columns from variable selection list whose values uniquely identify each row in the new table. If the Col ID variable sufficiently defines the split, no Group variables are required. Click **Group**, or drag the selection to the Group box.

Output Table Name:
  Optionally, you can specify a table name to use instead of the default name, **Untitled**.

The **Remaining Cols** radio buttons let you choose columns to include in the new table:

- **Keep All** includes all columns from the original table.
- **Drop All** removes all variables in the original table not specified for split functions.
- **Select...** displays a dialog for you to specify a subset of variables from the original table you want to keep in the new table with those used for splitting functions.

{ Tables }

The **Transpose** command creates a new JMP table that is the *transpose* of the active data table. The columns of the original (active) table are the rows of the new table, and the original table rows are the new table's columns. The new table has an additional column called labels, whose values are the column names of the active table. If there is a label variable in the active table, the values of that column are column names in the new transposed table. If there is no label column, the column names in the transposed table are Row1, Row2, ...Row*n* , where *n* is the number of rows in the original table. The **Transpose** facility has the following properties:

- The columns of the original table must be either all character or all numeric, except for a Label column and for columns in a summary table used for grouping.
- **Transpose** can transpose any selected subset of rows.
- **Transpose** can transpose groups of rows. For example, subsets defined by a summary table created with the **Group/Summary** command in the **Tables** menu (described above) transpose independently and stack to form a new transposed table.

The **Concatenate** command appends two or more tables end to end. The arrangement of the new data table depends on the columns in the original tables. **Concatenate** creates one column in the new table for each unique column name in all appended tables. Column names that are the same in multiple tables stack into a single column.

The Concatenate dialog in **Figure 2.16** lists all the open JMP data tables in the table selection list. Select the tables you want to concatenate, and click **Add** to build a list of tables in the list on the right. When you finish adding data table names, click **Concat**. This creates a new untitled data table that consists of all rows in the first selected table followed by all rows from the second table, and so on.

Figure 2.16 The Concatenate Dialog

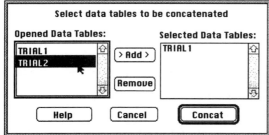

The **Join** command creates a new data table by joining two tables side by side. Tables can be joined

- by row number
- by matching the values in one or more columns that exist in both data tables
- in a Cartesian fashion where all values in a column of one data table are merged with all values in a column of another table.

Figure 2.17 The Join Dialog

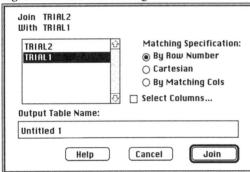

When you select the **Join** command, the dialog in **Figure 2.17** displays the name of the active table next to the word **Join**. You select an open data table to join **With** from the table selection list.

Click the radio buttons to use any of the following **Matching Specification** options:

⦿ **By Row Number**
Joining by row number joins tables side by side, matching them by row number. If the tables don't have the same number of rows, the columns from the shorter table have extra cells with missing values.

⦿ **Cartesian**
If you choose the Cartesian join, each row in the **Join** table joins with every row of the **With** table. The number of rows in the resulting table is the product of the rows in the two original tables.

⦿ **By Matching Cols...**
To join rows only when column values match, click **By Matching Cols** and complete the dialog that appears.

If you click **By Matching Cols** in the Join dialog (see **Figure 2.17**), the dialog shown in **Figure 2.18** then prompts you to select (highlight) a column in each table selection list whose values must match to complete a join. After selecting a column from each list, click **Match**. This displays the columns in the lower boxes. Select additional pairs of columns as needed. If you want to remove columns, select the same number of columns from each of the lower lists and click **Remove**.

Figure 2.18 The By Matching Columns Dialog

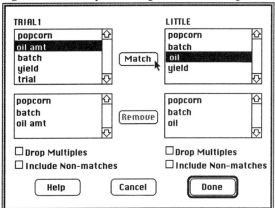

Note that matching columns do not require the same names and do not have to be in the same relative column position in both tables. When you click **Done**, the first column in the left-hand list pairs with the column selected in the right-hand list. Likewise, the second columns are paired, and so on. Rows are joined only if values match for all the column pairs.

After you choose columns whose values must match, additional matching options show for both tables. If you do not select any of these options, a Cartesian join occurs within each group of matching column values. Matching columns has the following two options:

Drop Multiples
You can specify **Drop Multiples** in either or both of the data tables being joined. If you drop multiples in both tables, only the first match is written to the new table. If you specify this option on only one table, the first match value joins with all matches in the other table.

Include Non-Matches
When you **Include Non-Matches** from a data table, the new data table includes each row from that data table even if there is no matching value. If there is no matching row in the other data table, cells for columns from the other data table have missing values for that row. You can specify this option for either or both data tables.

If you don't check **Select Columns** in the Join dialog (see **Figure 2.17**), all columns from both tables are included in the joined table. If you check **Select Columns**, the dialog in **Figure 2.19** prompts you to select columns from the original tables to include in the new table. If a name is unique, it is written directly to the new table. Columns with the same names in both tables are not overwritten. The new table includes both columns, named by appending the original column name to its data table name.

Figure 2.19 The Select Columns Dialog

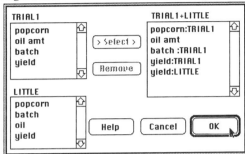

See Chapter 4, "The Tables Menu," for examples of the **Join** command.

The **Table Info** command displays the dialog shown in **Figure 2.20**. You can use this dialog to make notes about the current data table or to examine its attributes.

You can **Lock** the data table if you have not modified it since you last saved it. If you have made changes, the **Lock** box is dimmed and the **Dirty** box is checked.

You can also use the Table Info dialog to create a new data table by simply entering a new table name. The new data table is a duplicate of the original table. The original data table remains as it was when last saved, and the new table is not saved until you select the **Save** or **Save As** command.

Figure 2.20 The Table Info Dialog

The Table Info dialog can also be accessed from an attributes table, described next.

The **Attributes** command creates a new table called an *attributes table* from the active data table, called its *source table*. An attributes table has a row for each column in its source table and a column for each type of column characteristic.

Figure 2.21 shows the POPCORN.JMP table and its corresponding attributes table. The Table Info button in the upper–left corner of the attributes table accesses the source Table Info dialog.

Figure 2.21 A JMP Table and Its Attributes Table

POPCORN.JMP					
5 Cols / 16 Rows	popcorn [N]	oil amt [N]	batch [N]	yield [C]	trial [C]
7	plain	lots	small	10.6	1
8	gourmet	lots	small	18.0	1
9	plain	little	large	8.8	2
10	gourmet	little	large	8.2	2

Attributes of POPCORN.JMP									
Table Info 9 Cols / 5 Rows	Name [N]	Type [L]	Measure [N]	FW	N Dec	Lock [C]	Source [C]	Validation [N]	Role [N]
1	popcorn	Character	Nominal	7	?	No	No Formula	None	None
2	oil amt	Character	Nominal	6	?	No	No Formula	None	None
3	batch	Character	Nominal	6	?	No	No Formula	None	None
4	yield	Numeric	Interval	6	1	No	No Formula	None	None
5	trial	Numeric	Interval	5	0	No	No Formula	None	None

You can modify the column characteristics of a source table by editing values in its corresponding attributes table row. Changing a column's characteristics by editing a row in an attributes table is the same as changing characteristics in the Column Info dialog for that column. The advantage of using an attributes table is that you can change the characteristics of many columns at the same time.

At any time while you are editing an attributes table, you can update its source table by selecting the **Update Source** command in the dollar ($) pop–up menu at the lower left of the attributes table. Other commands in the dollar menu are as follows:

Sort Attributes
   accesses the Sort dialog used by the **Sort** command. You can sort the attributes table by any of its variables. When you update the source table. the columns rearrange according to the sort you requested.

Add Col Notes
   adds a character column to the attributes table called Col Notes. By default, it has a length of 64. You can change the length of this column using its Column Info dialog. If there are any currently existing column notes, they appear in the new Col Notes column. Any entry you make into the Col Notes column appears in the Column Info dialog for the corresponding source table column after you update it.

Max Col Note Width
   lets you specify a maximum width to allow for each column's notes, up to a maximum width of 127.

See Chapter 3, "JMP Data Tables," and Chapter 4, "The Tables menu," for more information about the **Attributes** command.

| Design Experiment... | | Design Experiment... |

Design Experiment accesses the Design of Experiments module in JMP. Design of experiments in JMP is documented in Chapters 25 through 30 of the *JMP Statistics and Graphics Guide*.

The Rows Menu

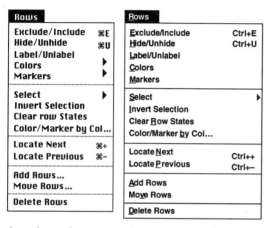

The first five commands in the **Rows** menu assign *row state* characteristics to selected rows. This section briefly describes row states. You can find more details about row states in Chapter 3, "JMP Data Tables." Row states can be saved permanently with the data table in a special row state column.

To select (highlight) a row, click the space in the row number area to the left of the row. This area is called the *row selection area*. To highlight multiple contiguous rows, drag down the row numbers or SHIFT–click the first and last rows of a desired range. To make a discontiguous selection, use COMMAND-click on the Macintosh or CONTROL–click under Windows to select the rows you want.

Rows are also highlighted (and selected) when you highlight points or bars in a corresponding graphical display. Because the JMP data table and any graphical display of that table are linked, selected points and rows always highlight simultaneously.

You can choose from a variety of **Rows** menu commands that affect highlighted rows such as
- excluding rows from further analysis
- hiding points in current graphical displays
- assigning special colors or markers for graphical display of points
- selecting all rows, or selecting rows assigned specific row state characteristics
- hiding points in current graphical displays
- automatic color and marker assignment
- adding or moving rows
- deleting rows.

To deselect all rows, click in the upper–left corner of the data table where the number of rows is shown. To deselect a single row, COMMAND–click that row's number (use CONTROL–click under Windows).

| Exclude/Include   ⌘E | | Exclude/Include   Ctrl+E |

Exclude/Include is a toggle command used to exclude selected rows from statistical analyses. Data remain excluded until you choose **Exclude/Include** is again for selected rows. Excluded rows are marked by the symbol shown to the left in **Figure 2.22**.

The **Exclude/Include** status is a row state characteristic. It can be saved permanently with the data table in a special row state column.

| Hide/Unhide   ⌘U | | Hide/Unhide   Ctrl+U |

Hide/Unhide is a toggle command that suppresses the display of points in all scatter plots. To hide data, highlight their points in any plot or in the spreadsheet and select **Hide/Unhide**. Data remain hidden until you choose **Hide/Unhide** again for selected rows. Hidden rows are marked by the symbol shown to the right in **Figure 2.22**.

The **Hide/Unhide** status is a row state characteristic. It can be saved permanently with the data table in a special row state column.

Figure 2.22 Excluded Rows and Hidden Rows

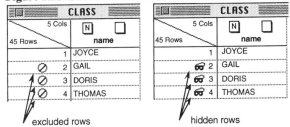

Warning➨ Hidden points can be included in statistical analyses even though they do not display in plots. Likewise, points can be excluded from an analysis but not hidden. These conditions could cause misleading plots and analyses.

| Label/Unlabel | | Label/Unlabel |

Label/Unlabel is a toggle command that labels points on all scatter plots. To label points, highlight them in any plot or in the spreadsheet, and select **Label/Unlabel**. By default, the row number is used as the label value on plots. However, if you designate a column in the spreadsheet as a Label column, its values show as labels in plots instead of the row numbers. To assign a column the Label role in the spreadsheet choose **Label** from the role assignment pop-up menu at the top of the column.

Data remain labeled until you choose **Label/Unlabel** again for selected rows. Labeled rows are marked by the tag shown in **Figure 2.23** The **Label/Unlabel** status is a row state characteristic. It can be saved permanently with the data table in a special row state column.

Figure 2.23 Labeled Rows

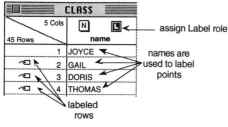

Colors changes highlighted points in all scatterplots to the colors you choose. To color a group of points, select the appropriate rows in the spreadsheet or select the points in a plot. Then choose a color from the **Colors** palette in the **Rows** menu. On the Macintosh you can *tear off* the color palette drag it anywhere on your desktop. To tear off the palette, click and drag it to the place you want it. Under Windows the **Show Colors Palette** command in the **Window** menu displays a floating colors palette that you can drag and place wherever you want. The floating colors palette is visible until you close it.

There are 65 colors including shades of gray for distinguishing data points in JMP. The default color is black. The **Colors** status is a row state characteristic. It can be saved permanently with the data table in a special row state column.

Markers assigns a plot character to replace the standard points in scatterplots and spinning plots. To assign one of the markers (shown to the right), select rows in the spreadsheet or select the corresponding points in a plot and choose a marker from the **Markers** submenu.

On the Macintosh you can *tear off* the markers palette and drag it anywhere on your desktop. To tear off the palette, click and drag it to the place you want it. Under Windows the **Show Markers Palette** command in the **Window** menu displays a floating markers palette that you can drag and place wherever you want. The floating markers palette remains visable until you close it.

There are eight markers for distinguishing data points in JMP. The default marker is a dot. The **Markers** status is a row state characteristic. It can be saved permanently with the data table in a special row state column.

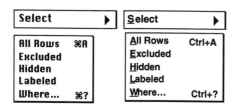

The **Select** command has a submenu with options for selecting all rows in a data table, or to select a subset of rows based on row states. Use the **Select** submenu as described next:

All Rows
> selects the entire JMP data table, the same as SHIFT–clicking the first and last rows in the spreadsheet. **Select Excluded** selects all excluded rows regardless of their current selection status and deselects all included rows.

Select Hidden
> selects all hidden rows regardless of their current selection status and deselects all rows that are not hidden.

Select Labeled
> selects all labeled rows regardless of their current selection status and deselects all unlabeled rows.

Where...
> lets you search for a specific value in a column and selects all rows where that value is found. **Where** displays the dialog shown in **Figure 2.24**, which prompts you to select a column, a comparison operation from the **Where** dialog pop–up menu, and a selection criterion value. When you click **OK**, **Where** highlights all rows that meet the search criterion.

Figure 2.24 The Select Where Dialog

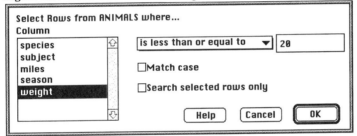

The **Where** command looks only at text strings. Before it completes the comparison, **Where** converts a numeric value to the character string as it appears in the table cell. The **Where** pop–up menu, shown to the right, lists the comparison operators. Multiple searches with the **Search selected rows only** box checked is the same as a logical And.

```
equals
does not equal
is greater than
is greater than or equal to
is less than
is less than or equal to
contains
does not contain
```

The **Invert Selection** command deselects all currently selected rows and selects all unselected rows.

| Clear Row States | Clear **R**ow States |

The **Clear Row States** command clears all active row states in the data table. All rows become included, visible, unlabeled, and show in plots as black dots. It does not affect row states saved in columns.

Color/Marker by Col...

The **Color/Marker by Col** command displays the dialog in **Figure 2.25**. JMP uses the levels of the variable you select to color or mark points in plots.

Figure 2.25 Color/Marker by Col Dialog

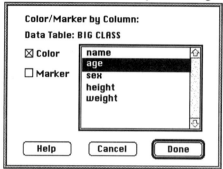

Locate Next

Locate Previous

The **Locate Next** command locates the first selected row after the *current row* and makes it blink briefly. The current row is the first row when you open a data table. The current row changes to the most recent row you edited or identified by a **Locate Next** or **Locate Previous** command. You can set the current row by OPTION–clicking anywhere in a row (ALT–click under Windows). Each time you choose the **Locate Next** command, the next selected row is found and blinks. A beep tells you when the last selected row is located.

The **Locate Previous** command behaves the same as **Locate Next** but locates the first selected row before the *current row* and makes it blink briefly.

The **Add Rows** command displays the dialog to the left in **Figure 2.26**. The Add Rows dialog prompts you to enter the number of rows to add and to specify their location in the table. Click the appropriate radio button to add rows at the beginning of the table (**At Start**), the end of the table (**At End**), or after a row number you specify (**After Row #**). The new rows appear in the table when you click **Add**, and have missing values that can be filled by typing or pasting in data.

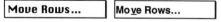

The **Move Rows** command operates on highlighted rows. It moves highlighted rows to the location specified by the dialog in **Figure 2.27**. Click the appropriate radio button to move highlighted rows to the beginning of the table (**To Beginning**), to the end of the table (**To End**), or after a row number you specify in the **After Row #** box. The rows appear in the new location when you click **Move**.

Figure 2.26 The Add Rows and Move Rows Dialogs

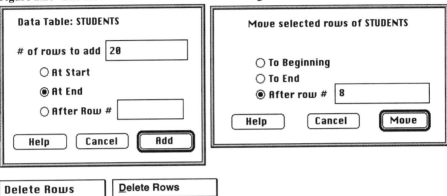

The **Delete Rows** command deletes all selected rows from a JMP data table. Use the **Undo** command on the **File** menu to undo an accidental deletion.

The Cols Menu

Cols	
Assign Roles...	
Clear All Roles	⌘K
New Column	
Add Columns...	
Move to First	⌘<
Move to Last	⌘>
Move Columns...	
Hide Columns	
Unhide...	
Column Info	⌘I
Delete Columns	

Cols	
Assign Roles...	
Clear All Roles	Ctrl+K
New Column...	
Add Columns...	
Move to First	Ctrl+ <
Move To Last	Ctrl+ >
Move Columns...	
Hide Columns	
Unhide...	
Column Info	Ctrl+I
Delete Columns	

Cols menu commands act on selected columns in the current data table. To select a column, click the background area above the column name. This area is called the *column selection area*. To highlight multiple columns, drag across their column selection areas or SHIFT–click the first and last columns of a desired range. Use COMMAND–click on the Macintosh or CONTROL–click under Windows to make a discontiguous selection.

The **Cols** menu commands let you

- assign or clear a column's analysis role
- create or insert new columns and access the JMP calculator to compute column values
- rearrange the order of columns
- hide columns temporarily
- change column characteristics
- delete unwanted columns.

| Assign Roles... | Assign Roles... |

The **Assign Roles** command displays the dialog shown in **Figure 2.27** for assigning the following analysis roles to columns:

- **X** identifies columns as independent or predictor variables that are regressors, model effects, or classification variables that divide the rows into sample groups.
- **Y** identifies columns as response or dependent variables whose distributions are to be studied.
- **Weight** identifies a column whose values supply weights for each response variable.
- **Freq** identifies a column whose values assign a frequency to each row for an analysis.
- **Label** identifies a column whose values specify labeling values for plotted points.

You can also specify variable roles by using the role assignment box at the top of each column in the spreadsheet or by selecting an analysis platform and responding to an assign roles prompt.

If you use the **Assign Roles** command or the role assignment boxes in the spreadsheet, your role assignments remain in effect until you specifically change them. If you assign a sufficient number of roles, the **Analyze** menu platforms use these assignments to determine the type of analysis to do, and automatically completes the analysis. If you haven't assigned roles or you choose an analysis that requires more roles than you provided, a dialog prompts you for more role assignment information. Role assignments in response to the specific analysis dialog affect only that analysis.

Figure 2.27 The Assign Roles Dialog

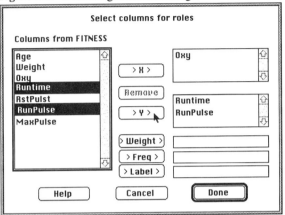

The **Clear All Roles** command removes role assignments from all variables in the active data table.

[New Column...]   New Column...

The **New Column** command lets you define one or more new columns by completing the New Column dialog shown in **Figure 2.28**. The dialog asks you to name the new column and provide column characteristics. If you need to compute values for a new column the **Formula** selection in the **Data Source** pop–up menu gives you access to the calculator when you click **OK**.

Figure 2.28 The New Column Dialog

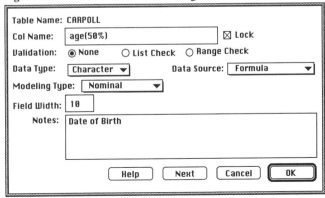

Use the **New Column** dialog elements as follows:

Table Name
displays the name of the active data table and cannot be changed.

Col Name
lets you type in the new column's name using up to 32 keyboard characters. If you choose a long name, you can expand the width of the column in the spreadsheet by dragging the column boundary with the mouse.

Lock
renders a column *uneditable*. If you use a formula to compute values for the column, it is automatically locked. Also, you can click in the **Lock** box to protect any column's values from modification.

Validation
lets you set up a table of acceptable values or an acceptable range of values for a column. The **List Check** radio button shows the dialog shown to the left in **Figure 2.29** to enter a list of valid values for a column. If you click the **Range Check** radio button, the dialog shown to the right lets you specify a range of values and range limit conditions.

Figure 2.29 Data Entry Validation Dialogs

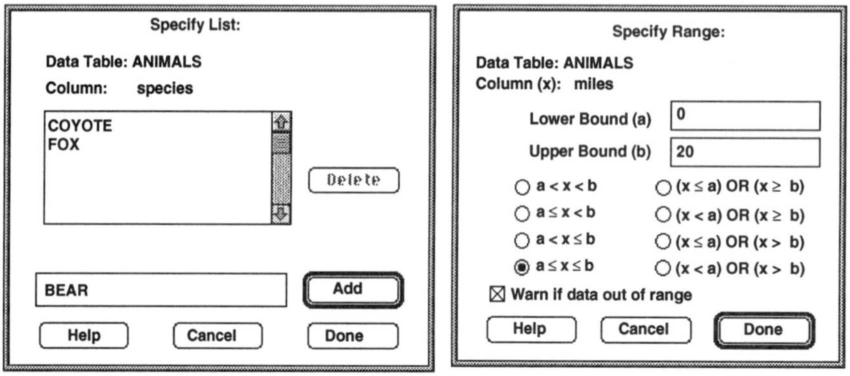

Data Type
assigns a data type to the new column according to your selection from its menu.

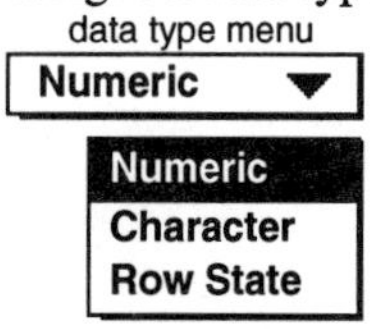

Numeric columns must contain numbers.

Character columns can contain any characters including numbers. In character columns, numbers are treated as discrete values instead of continuous values.

Row State columns contain special information that affects the appearance of graphical displays. Row states include color, marker, selection status (highlighted or not highlighted), include/exclude status, hide/unhide status, and label/unlabel status.

Active row state information displays in the row selection area of the spreadsheet. You can copy or add row state information to or from a special row state column using pop-up commands located at the top of the column as illustrated next.

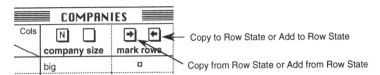

Data Source

tells the source of the column values:

- **No Formula** values are imported, pasted, or keyed into a column. These values are editable.

- **Formula** specifies that values of a column are calculations. You use the column's calculator window to construct a formula that computes these values. The formula can include existing columns, constants, conditional logic, and a variety of functions. To see a column's calculator, choose **Formula** from the pop-up menu and click **OK**.

  A column of computed values is locked and cannot be edited. However, you can disassociate a column from its formula by selecting **No Formula** as the data source in its Column Info dialog. For calculator details see Chapter 5, "Calculator Functions," and Chapter 6, "Using the Calculator."

- **Instrument** values are transmitted to a Macintosh with a measuring instrument and the Macintosh Communication Toolbox. The toolbox lets you connect your Macintosh to a measuring device. The **Instrument** data source tells JMP to read measures as they are taken and update a data table and control charts. This feature is not currently available under Windows.

Note → If you do not already have the Macintosh Communication Toolbox, contact the JMP Sales Department.

Modeling Type

specifies the way you want JMP to use column values. The modeling type is a column characteristic that JMP uses to determine how to analyze data. You can change a column's modeling type as long as it corresponds to one of the following data types:

- **Continuous** columns must contain numeric values.

- **Ordinal** columns can contain characters or numbers. The analysis platforms treat ordinal values as discrete categorical values that have an order. If the values are numbers, the order is the numeric magnitude. If the values are character, the order is the collating sequence.

- **Nominal** columns can contain either numeric or character values. All values are treated by analysis platforms as though they are discrete values.

Field Width

specifies the field width needed to accommodate the largest number of digits or characters you plan to enter in the new column. The maximum field width is 40 for numeric values and is 255 for character values.

Format
specifies a format to display numeric column values:
- **Fixed Decimal** displays a value rounded to the number of decimals you specify. This is especially useful for showing dollars and cents amounts.
- **Best** is the default format. This means JMP considers the precision of the values and chooses the best display for them.
- **Date & Time** assumes the numeric value in a column represents the *number of seconds since midnight, January 1, 1904*, and displays a corresponding date format. **Date & Time** has an additional pop–up menu for choosing a specific date representation. The following examples show formats for the date December 31, 1995 (its unformatted value is 2903212800).

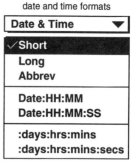

The **Short** date format displays a date as *dd/mm/yy*, giving 12/31/95.

The **long** date format displays a date value as *weekday, month day, year*, giving Sunday, December 31, 1995.

The **Abbrev** date format display is the same as the **long** format except that weekday and month have three–character abbreviations, giving Sun, Dec 31, 1995.

The Date:HH:MM and **Date:HH:MM:SS** formats display a date value as a **Short** date followed by the number of hours, minutes, and seconds after midnight of that date. This example has no hours or minutes. Its formatted values are 12/31/95 12:00 AM and 12/31/95 12:00:00 AM.

The *:***Days:HH:MM** and *:***Days:HH:MM:SS** formats show the number of days, hours, Minutes, and seconds since January 1, 1904. The results for December 31, 1995 are :33602:0:0 and :33602:0:0:0.

Note → The decimal points for the fixed decimal and best formats, month and day names, the date field separators, and the order of the date elements depend on the localized system software in your country. Also, if you are using the Macintosh operating System 7.1 or later, JMP uses the date forms you define in the Date and Time control panel for Long, Short, and Abbreviated.

See Chapter 3, "JMP Data Tables," for details about characteristics of data.

Notes
is a text entry area you can use to document information about a new column. The text entry can contain up to 255 characters.

Next
creates a new column with the current New Column dialog information, and makes the dialog ready to accept information for another new column.

Chapter 2 *The Menu Bar* 47

{ Cols }

Add Columns...		Add Columns...

The **Add Columns** command displays the dialog in **Figure 2.30**, which lets you add more than one column at a time to a table. You specify the number of columns to add, their location, field width, and type. Check the appropriate radio button to add columns at the beginning of the table (**At Start**), after the rightmost column (**At End**) or inserted between columns (**After Selected Col**).

Figure 2.30 The Add Columns Dialog

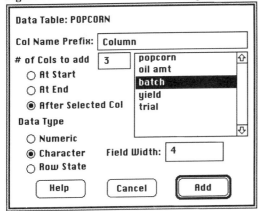

By default, the new column names Column 1, Column 2, and so forth. You can enter any value in the **Col Name Prefix** box and that prefix is used in place of Column. Other column characteristics are the same for all the new columns. You can change the column names and characteristics by editing them in the spreadsheet, in an attributes table, or in the Column Info dialog. See the New Column command discussed previously for more information about specifying column characteristics.

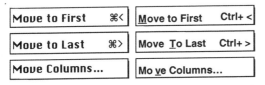

Move to First, **Move to Last**, and **Move Cols** operate on highlighted rows. **Move to First** moves columns to the left–most position on the spreadsheet. **Move to Last** moves columns to the right–most position. **Move Cols** displays the dialog shown in **Figure 2.31** and moves columns to the location you specify. If you mistakenly move one or more columns, use the **Undo** command in the **Edit** menu to restore the previous order.

Figure 2.31 The Move Columns Dialog

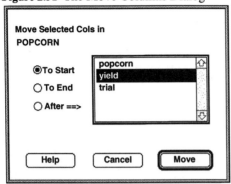

The **Hide** command hides selected column(s) in the spreadsheet but does not remove them from the data table. Use the **Unhide** command to reshow hidden columns.

Note➡ The number of columns showing in the upper–left corner of the data table does not change when you hide columns. Hidden columns are not apparent except to note that the **Unhide** command is not dim.

The **Unhide** command displays a dialog that lists all hidden column (**Figure 2.32**). You can select any subset of the hidden columns to reshow on the spreadsheet. When you unhide a column, it becomes the last column in the spreadsheet.

Figure 2.32 The Unhide Dialog

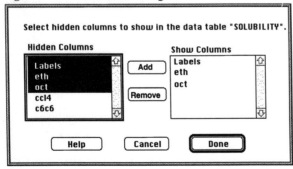

Chapter 2 The Menu Bar   49

```
Cols
```

| Column Info   ⌘ I | Column Info   Ctrl+ I |

The **Column Info** command displays the dialog used by the **New Column** command (see **Figure 2.28**), except there is no **Next** button for adding new columns. You can use the Column Info dialog at any time to change a single column's attributes. For example, you can add data validation to a column if you click a data validation radio button as shown in **Figure 2.33**.

If you select **Formula** in the **Data Source** pop–menu for a column that has **No Formula** as its data source, the calculator window opens after you click **OK**. If you click the picture of an existing formula (at the lower–left corner in the Column Info dialog), the calculator window opens so you can modify the formula.

Figure 2.33 The Column Info Dialog

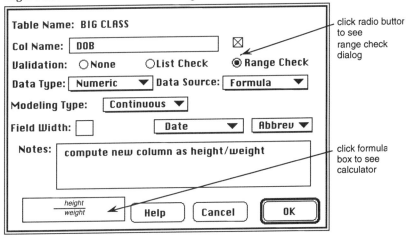

| Delete Columns | De<u>l</u>ete Columns |

The **Delete Columns** command removes the selected columns from the data table. If you accidentally remove columns, you can use the **Undo** command in the **Edit** menu to restore them.

The Analyze Menu

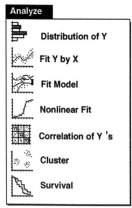

The **Analyze** menu is the heart and soul of JMP. Each Analyze patform produces a window that shows statistical text reports with supporting graphical displays.

Statistical platforms require variable role assignments. You assign roles with the role assignment pop–up menu at the top of each column in the spreadsheet or with the **Assign Roles** command in the **Rows** menu. If roles were not previously assigned, the **Analyze** platforms prompt you to assign them.

In addition to the menu of commands, the following set of tools shows in the toolbar under Windows.

Each **Analyze** command is briefly discussed here. See the *JMP Statistics and Graphics Guide* for more description of each command, its options, and comprehensive examples.

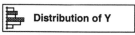

Distribution of Y describes a distribution of values with histograms and other graphical and textual reports:

- Continuous columns display a histogram, an outlier box plot, and a quantiles box plot. You can change the width of the histogram bars using the hand tool from the **Tools** menu.
- Nominal or ordinal columns are shown with a histogram of relative frequency for each level of the ordinal or nominal variable and a mosaic (stacked) bar chart.

Text reports support each of the distribution plots. The reports show selected quantiles and moments of continuous values. Tables of counts and proportions support nominal and ordinal values.

Save commands let you save information such as rank, level number, standardized values, and other statistics as new columns in the data table.

Chapter 2, "Distributions" in the *JMP Statistics and Graphics Guide* describes the **Distribution of Y** command in detail and gives examples.

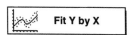

Fit Y by X studies the relationship of two variables. This platform shows plots with accompanying analyses for each pair of X and Y variables. The kind of analysis done depends on the modeling types (continuous, nominal, or ordinal) of the X and Y columns. **Figure 2.34** illustrates each type of plot produced by the different combinations of X and Y modeling types.

Figure 2.34 Fit Y by X Platforms

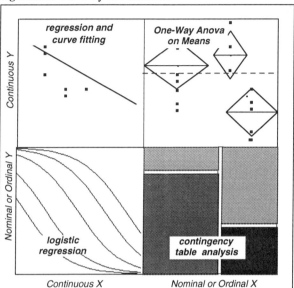

- If both X and Y have continuous modeling types, **Fit Y by X** displays a scatterplot. Using options you can explore various regression fits for the data and choose the most suitable fit for further analysis. Each fit is accompanied by tables with supporting statistical analyses and parameter estimates.
- If X is nominal or ordinal and Y is continuous **Fit Y by X** plots the distribution of Y values for each discrete value of X. You can use pop–up menu options to see a means diamonds and a box plot for each value of X, and to compare group means with comparison circles. An accompanying text report shows a one–way analysis of variance table. Optionally, you can request nonparametric analyses, see multiple comparisons, and test homogeneity of variance.
- If X has continuous values and Y has nominal or ordinal values, **Fit Y by X** performs a logistic regression and displays a family of logistic probability curves. Accompanying tables show the log likelihood analysis and parameter estimates for each curve.

  Note➡ Logistic regressions of ordinal columns are parameterized differently from logistic regressions of nominal columns and therefore produce different results.
- If both X and Y are nominal or ordinal values, a contingency table mosaic chart is shown. A mosaic chart consists of side–by–side divided bars for each level of X. Each bar is divided into proportional segments representing the discrete Y values. Accompanying tables show statistical tests, and frequency, proportion, and chi–square values for each cell. Optionally, you can request a correspondence analysis.

Individual chapters in the *JMP Statistics and Graphics Guide* describe each type of analysis given by the **Fit Y by X** command.

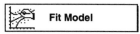

Fit Model

Fit Model displays the dialog shown in **Figure 2.35**. This dialog lets you tailor an analysis using a model specific for your data. The variable selector list in the dialog lists all columns in the current JMP data table. You select columns, assign roles, and choose the model to fit.

Figure 2.35 The Fit Model Dialog

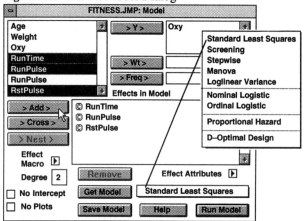

Fit model performs a fit of one or more Y variables by the X variables selected. You can select the kind of model appropriate to your data from the pop–up menu of fitting *personalities* shown in **Figure 2.35**. The fitting personalities available depend on the kind of responses you select. The following list briefly describes the different fitting techniques:

- **Standard Least Squares** gives a least squares fit for a single continuous response, accompanied by leverage plots and an analysis of variance table.
- **Screening** produces an exploratory screening analysis for single or multiple Y columns with continuous values.
- **Stepwise** gives a stepwise regression for a single continuous Y and all types of effects.
- **Manova** performs a multivariate analysis of variance for multiple continuous response columns. **Manova** displays a dialog that lets you fit multivariate models interactively.
- **Loglinear Variance** is for a single continuous response and estimates parameters that optimize both a mean and a variance.
- **Nominal Logistic** for a single nominal response, does a nominal regression by maximum likelihood.
- **Ordinal Logistic** for a single ordinal response, does an ordinal cumulative logistic regression by maximum likelihood.
- **Proportional Hazard** performs a proportional hazard model fit for survival analysis of censored data with a single continuous response.

- **D–Optimal Design** does a D–optimal design search from a candidate data table if you do not specify a Y variable.

Individual chapters in the *JMP Statistics and Graphics Guide* document each technique offered by the model fitting platform.

Nonlinear Fit fits nonlinear models, which are models that are nonlinear in their parameters. The **Fit Nonlinear** command launches an interactive fitting facility. You orchestrate the fitting process as a coordination of three important parts of JMP: the data table, the calculator, and the nonlinear fitting platform.

You define the nonlinear prediction formula with the calculator and launch the nonlinear fit platform with the response variable in the Y role, and the model column with its fitting formula in the X role. You interact with the platform through the *Nonlinear Fitting Control Panel*. The control panel has

- buttons to start, stop, and step through the control the fitting process, and to reset parameter values
- fitting options to specify loss functions and computational methods
- a processing messages area
- a list of current and limit convergence criteria and step counts, current parameter estimates, and error sum of squares
- options to specify the alpha level for confidence intervals and delta for numerical derivatives.

The nonlinear platform can show the model and the derivatives of the model with respect to each of its parameters, and the fitting solution reports. There are features that give confidence intervals on the parameters and plot the resulting function if it is of a single variable. You can also save the SSE values in a data table with a grid for plotting them.

Chapter 14, "Nonlinear Regression" in the *JMP Statistics and Graphics Guide* describes the **Nonlinear Fit** command in detail and gives examples.

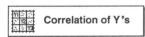

Correlation of Y's explores how multiple variables relate to each other, and how points fit that relationship. This platform helps you see correlations between two or more response (Y) variables, look for points that are outliers, examine principal components to look for factors.

The Correlation of Y's platform appears showing a correlation matrix. Options show
- the inverse correlation matrix
- a partial correlation matrix
- pairwise correlations with accompanying bar chart

- nonparametric correlations
- a matrix of bivariate scatterplots with a plot for each pair of Y variables.

Correlation of Y's also shows a graphical outlier analysis that includes
- a Mahalanobis distance outlier plot
- a jackknifed multivariate distance outlier plot where the distance for each point is calculated excluding the point itself.

There are options with these plots to save the distance scores.

You can also request principal components, standardized principal components, rotation of a specified number of components, and factor analysis information.

Chapter 15, "Correlations and Multivariate Techniques" in the *JMP Statistics and Graphics Guide* describes the **Correlation of Y's** command.

 **Cluster**

The **Cluster** command clusters rows of a JMP table. You can choose either a hierarchical or a K–means clustering method .

The hierarchical cluster platform displays results as a tree diagram of the clusters called a dendogram followed by a plot of the distances between clusters. The dendogram has a sliding cluster selector that lets you identify the rows in any size cluster. You can use options to save the cluster number of each row. **Cluster** (hierarchical) uses these five clustering methods:

- **Average** linkage computes the distance between two clusters as the average distance between pairs of observations, one in each cluster.
- **Centroid method** which computes the distance between two clusters as the squared Euclidean distance between their means.
- **Ward's minimum variance** method uses the distance between two clusters as the Anova sum of squares between the two clusters added up over all the variables.
- **Single linkage** uses the distance between two clusters that is the minimum distance between an observation in one cluster and an observation in the other cluster.
- **Complete linkage** uses the distance between two clusters that is the maximum distance between an observation in one cluster and an observation in the other cluster.

The k–means clustering approach finds disjoint clusters on the basis of Euclidean distances computed from one or more quantitative variables. Every observation belongs to only one cluster—the clusters do not form a tree structure as with hierarchical clustering. You specify the number of clusters you want.

Chapter 16, "Cluster Analysis" in the *JMP Statistics and Graphics Guide* describes the **Cluster** command in detail and shows clustering examples.

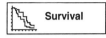 **Survival**

The **Survival** command lets you analyze survival data three ways:

- product–limit (Kaplan–Meier) life table survival computations with estimation of Wiebull, log normal and exponential parameters.
- regression analysis that tests the fit an exponential, a Weibull, or a log normal distribution
- proportional hazard regression analysis that fits a Cox model.

Note→ You can also use the Nonlinear Fit platform to handle nonlinear models with loss functions for other parametric survival modeling.

Chapter 17, "Survival Analysis" in the *JMP Statistics and Graphics Guide* describes the **Survival** command in detail.

The Graph Menu

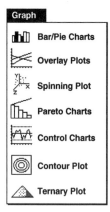

Graph menu commands produce windows that contain specialized graphs or plots with supporting tables and statistics.

The Graph platforms require role assignments. You can assign roles with the pop–up menu at the top of each spreadsheet column or with the **Assign Roles** command in the **Rows** menu. If roles were not previously assigned, the **Graph** platforms prompt you to assign them.

Note ➡ Graphs produced by the **Bar/Pie Charts, Overlay Plot**, and **Control Charts** commands update automatically when you add rows to the current data table and then click the chart or graph to activate it. You can add rows yourself or receive data from an external source using a measuring instrument and the Macintosh Communication Toolbox.

If you are using Windows, the following set of graph tools appear in the toolbar.

For a detailed discussion of each **Graph** menu command see the *JMP Statistics and Graphics Guide*.

The **Bar/Pie Charts** command gives a chart for every numeric Y specified. The Y's are the statistics you want to chart. Initially, you see a vertical bar chart, but options let you show horizontal bar charts, line charts, step charts, needle charts, or pie charts.

Bar/Pie Charts assumes the data are summarized, giving a unique set of values for the X variables you specify. If multiple X variables have the same values, the chart facility assumes your data have not been summarized. It advises you to summarize them and displays the **Group/Summary** command dialog. When you complete the summary dialog and click **Done**, the **Chart** command continues, and displays the default bar chart.

You can specify up to two X variables for grouping on the chart itself. The first X is the group variable, and the second X is the level (subgroup) variable. If you do not specify an X variable, then each row is a bar. The X variables do not have to be sorted, but each combination of the X's must yield a distinct category. Groups and levels display in the order that they occur in the data table (see the example in **Figure 2.36**).

Figure 2.36 Example Bar Chart of Means

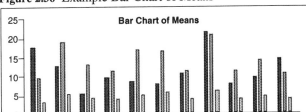

Chapter 18, "Bar, Line, and Pie Charts," in the *JMP Statistics and Graphics Guide* describes the **Bar/Pie Charts** command and shows examples.

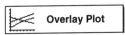 **Overlay Plot**

The **Overlay Plot** command overlays a plot of a single numeric X column and all numeric Y variables. The axis can have either a linear or a log scale Optionally, the plots for each Y can be shown separately with or without a common X axis.

By default, the values of the X variable are sorted in ascending order, and the points are plotted and connected in that order. You have the option of plotting the X values as they are encountered in the data table.

If the given X variable is not numeric, the **Overlay Plots** command calls **Bar/Pie Charts** described above, which displays an overlaid line chart.

Note➡ If you want scatterplots of two variables at a time with regression fitting options, use the **Fit Y by X** command instead of **Overlay Plot.**

Figure 2.37 Example of Overlay Line Plot

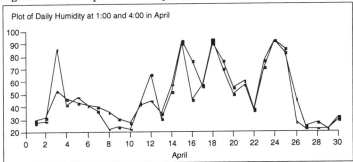

Chapter 19, "Overlay Plots," in The *JMP Statistics and Graphics Guide* describes the **Overly Plot** command in detail and shows examples of plotting data.

 **Spinning Plot**

Spinning Plot produces a three–dimensional spinable display of values from any three numeric columns in the data table as shown in **Figure 2.38**. It also produces an approximation to higher dimensions through principal components, standardized principal components, rotated components, and biplots. Options let you save principal component scores, standardized scores, and rotated scores.

The Spinning Plot platform also gives factor–analysis–style rotations of the principal components to form orthogonal combinations that correspond to directions of variable clusters in the space. The method used is called a *varimax rotation,* and is the same method that is traditionally used in factor analysis.

Figure 2.38 Example Spinning Plot

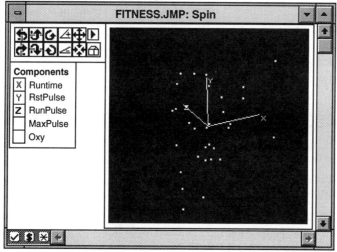

Chapter 20, "Three–Dimensional Viewing," in the *JMP Statistics and Graphics Guide* describes the **Spinning Plot** command in detail and shows examples of plotting data and computing principal components .

 **Pareto Chart**

The **Pareto chart** command gives charts that display counts or the relative frequency of problems in a quality–related process or operation. Pareto charts compare quality–related measures or counts in a process or operation. The defining characteristic of pareto charts is that the bars are in descending order of values, which visually emphasizes the most important measures or frequencies.

When you select **Pareto Chart**, a variable selection dialog prompts you to assign variable roles. **Pareto Chart** uses a single Y variable, called a process variable, and gives
- a simple Pareto chart when you do not specify an X (classification) variable
- a one–way comparative Pareto chart when you specify a single X variable.
- a two–way comparative chart when there are two X variables.

The Pareto chart facility does not distinguish between numeric and character variables, or between modeling types. All values are treated as discrete, and bars represent either counts or percentages.

Figure 2.39 Example One–way Comparative Pareto Chart

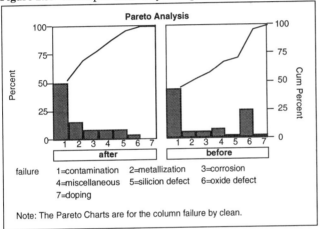

Chapter 21, "Pareto Charts," in the *JMP Statistics and Graphics Guide* describes the **Pareto Chart** command in detail.

Control Charts

The **Control Charts** command creates dynamic plots of sample subgroups as they are received and recorded. Control charts are a graphical analytic tool used for statistical quality improvement. Control charts can be broadly classified according to the type of data analyzed:
- Control charts for *variables* are used when the quality characteristic to be analyzed is measured on a continuous scale.
- Control charts for *attributes* are used when the quality characteristic is measured by counting the number of nonconformities (defects) in an item or by counting the number of nonconforming (defective) items in a sample.

The concepts underlying the control chart are that the natural variability in any process can be quantified with a set of control limits, and that variation exceeding these limits signals a special cause of variation. In industry, control charts are commonly used for studying the variation in output from a manufacturing process. They are typically used to distinguish variation due to *special* causes from variation due to *common* causes.

Figure 2.40 Example of Mean Control Chart Showing Zones

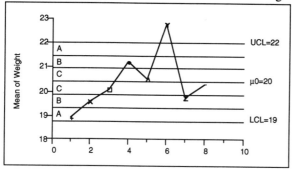

The Control Chart platform offers the following kinds of charts:
- mean, range, and standard deviation
- individual measurement and moving range
- P, NP, C, and U
- UWMA and EWMA
- cusum.

Chapter 22, "Statistical Control Charts," in the *JMP Statistics and Graphics Guide* describes the **Control Charts** command in detail.

 **Contour Plot**

The **Contour Plot** command constructs a contour plot from X and Y values for a response variable, Z. **Contour Plot** assumes the X and Y values lie in a rectangular coordinate system, but the observed points do not have to form a grid.

The plots in **Figure 2.41** show levels of electrical brain activity. Some of the Contour platform options are
- show or hide data points
- show or hide triangulation and boundary
- specification of levels
- show a line contour or fill areas

Figure 2.41 Example Contour Plots without Area Fill

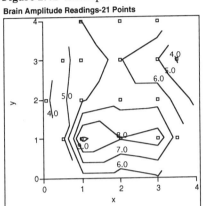 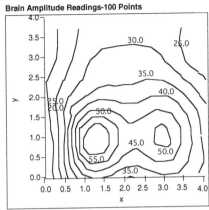

Chapter 23, "Contour Plots," in the *JMP Statistics and Graphics Guide* describes the **Contour Plot** command in detail and shows examples of plotting data.

 **Ternary Plot**

The **Ternary** command constructs a plot using triangular coordinates. **Figure 2.42** shows examples of Ternary plots. The Ternary platform uses the same options as the Contour platform for building and filling contours. In addition it has specialized tools in the **Tools** menu that let you adjust the axes with the hand tool and read the triangular axes values with the crosshair tool.

Figure 2.42 Example Ternary Plots Showing Grid Lines and Contours

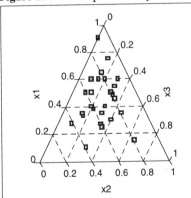 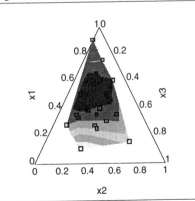

Chapter 24, "Ternary Plots," in the *JMP Statistics and Graphics Guide* describes the **Ternary Plot** command in detail and shows ternary plot examples.

The Tools Menu

The **Tools** menu is a palette of special tools that determine the effect of mouse actions. The default tool is the arrow. The function of the arrow is territorial. For example,

- when the arrow is in a report window, it reveals text reports, accesses pop–up menus, displays point labels, and highlights histogram bars
- when used in a spreadsheet, the arrow accesses pop–up menus at the top of each column, selects rows and columns, and selects text for editing.

Other tools cut or copy all or part of a report, access help windows, manipulate graphical displays, and highlight points of interest in a plot. These tools function in report windows but not on the spreadsheet.

← Note: If you are using Windows, this set of tools appears in the Windows toolbar.

On the Macintosh you can *tear off* the Tools menu and drag it anywhere on your desktop. To tear off the **Tools** menu, click and drag it to the place you want it. Under Windows the **Show Tools Palette** command in the **Window** menu displays a floating tools palette to drag and place wherever you want.

The hand tool (or grabber) is for direct manipulation or *grabbing* in plots and charts. In a text report, the hand tool behaves the same as the arrow tool.

The hand behaves in graphs and plots as follows:

On histograms
: you can use the grabber tool to change the number of bars in a histogram or to shift the boundaries of the bars on the axis.

On Spinning Plots
: the grabber tool spins the plot. To spin a plot, grab the plot with the hand by holding down the mouse button, and then move the hand about.

In a scatterplot matrix
: the grabber can drag the position of a column of scatterplots to a new position in the matrix.

In a cusum control chart
: the grabber relocates the origin of the V–mask.

In a Ternary plot
: moving the hand tool zooms in on triangular areas.

The question mark icon accesses JMP Help. Select the help tool and then click graphs, plots, or tables to see help windows.

Each help window persists as long as the cursor is within the window. You can shift–click the question mark to bring up a help window, which persists until you close it. Some help windows have buttons to reveal further help details and persist automatically.

The brush tool is for highlighting an area of points in plots. When you click, a rectangle appears. Move the rectangle over points to highlight them in the plot and in the active data table. SHIFT-click to extend the selection. OPTION-click (ALT–click under Windows) to change the size of the selection rectangle and also extend the selection. If the brush tool is not in a plot area, it behaves the same as the arrow tool.

The crosshair is a movable set of axes used to measure points and distances in graphical displays. For example, use the crosshair on a fitted line or curve to identify the response value for any predicted value. The values where the crosshair intersects the vertical and horizontal axis appear automatically as you drag the crosshair within a plot. On a ternary plot, this tool displays triangular crosshair lines.

The scissors are for selecting cut–and–paste territory from a report. Drag the scissors diagonally over any part of the report to define a rectangular area you want to copy to the clipboard. Hold down the OPTION key (ALT key under windows) and drag the scissors to select blocks of the report window. Use SHIFT-click and drag the scissors to select discontiguous sections of report areas.

The lasso tool lets you highlight an irregular area of points in plots. Drag the lasso around any set of points. When you release the lasso it automatically closes and highlights the points within the enclosed area. Use SHIFT-lasso to drag the lasso around discontiguous irregular areas of points.

The magnifying glass tool automatically zooms in on any area of a plot. When you click the magnifier, the point or area where you click becomes the center of a new view of the data. The scale of the new view is enlarged approximately 25%, giving you a closer look at interesting points or patterns. Use OPTION-click on the Macintosh or ALT-click under Windows at any time to restore the original plot. On a ternary plot you can drag the magnifier tool to zoom the triangular axes.

The annotate tool places a text box wherever you click in a JMP window. You can key in notes and remove them at a later time, draw lines to make a special point, or use the annotate tool to enhance a JMP graphical display.

See Chapter 7, "Report Windows," for examples of using JMP tools.

The Window Menu

The **Window** menu helps you organize the windows produced during a JMP session. All open windows generated in a session are listed below the last line in the **Window** menu. In the menus shown to the left, the data table called BIGCLASS.JMP (BIG CLASS on the Macintosh), a **Distribution of Y** analysis window for the BIG CLASS table, and an Untitled window are open; the **Distribution of Y** window active; when you click a window name, it becomes the active window.

The **Window** menu under Microsoft Windows is different than Macintosh **Window** menu. The first six commands (described next) are not in the Macintosh **Window** menu.

Show Markers Palette

The **Show Markers Palette** opens a window that displays the markers palette given by the **Markers** command in the Rows menu. You can drag the Markers window to any convenient location on your desktop and it remains open until you close it. This command serves the same function as *tearing off* the markers palette on the Macintosh.

Show Colors Palette

The **Show Colors Palette** opens a window that displays the colors palette given by the Colors command in the **Rows** menu. You can drag the Colors window to any convenient location on your desktop and it remains open until you close it. This command serves the same function as *tearing off* the colors palette on the Macintosh.

Show Tools Palette

The Show Tools Palette opens a window that displays the **Tools** Menu and displays the JMP tools as a palette on your desktop. You can drag the Tools window to any convenient location on your desktop and it remains open until you close it. This command serves the same function as *tearing off* the **Tools** Menu on the Macintosh.

Cascade

The **Cascade** command arranges open windows side by side so all of them are visible.

Tile

The **Tile** command arranges open windows so that the title bar of each window is visible.

Arrange Icons

The **Arrange Icons** arranges all program–item icons for a selected group into rows. Or, if a group icon is selected, **Arrange Icons** arranges all group icons into rows.

Note→ The remaining commands in the **Windows** menu are the same on the Macintosh and under Microsoft Windows.

Redraw ⌘D | Redraw Ctrl+D

The **Redraw** command redraws the active window. It is useful for cleaning up both spreadsheet views and graphical displays that have accumulated stray imperfections resulting from high–speed, dynamic handling of windows.

Move To Back | Move to Back

The **Move To Back** command moves the active window behind all other windows generated by the current JMP session, leaving the next window in the sequence showing.

Hide | Hide

The **Hide** command suppresses the display of the active window but does not close it. To reshow a hidden window, select the window name from the list beneath the dotted line in the **Window** menu.

New data View | New Data View

The **New data View** command displays a new spreadsheet view of an open data table. The new view is linked to the original view and all corresponding analysis windows. Changes made to a new view reflect on the original view when it is made active.

Set Window Name... | Set Window Name...

The **Set Window Name** command lets you change the name of any active JMP window. This is especially useful if you generate multiple untitled windows and need to distinguish among them during the JMP session.

Close *type* Windows | Close *type* Windows

The **Close *Type* Windows** command closes all windows of a given type, where the active window determines the *Type*. For example, if a data table window is active, this command becomes **Close Data Windows**, and it closes all data tables when you select it.

The Help Menu (Microsoft Windows)

Help
Contents...
Search for Help On...
Statistical Guide
How to Use Help...
About JMP...

The Windows version of JMP has the standard Windows **Help** menu. Windows users will find the menu items similar to those in other applications.

For details about the Windows Help menu in general, consult your Microsoft Windows User's Guide.

See Chapter 7, "JMP Report Windows," for more information about using JMP help.

Contents

The **Contents** command displays the same help access window that the Macintosh gives with the **About JMP** command in its Apple menu (see **Figure 2.43**). This is the topmost help window. This main JMP help window has buttons for further help. All the help documents are arranged hierarchically with buttons for accessing more detailed information.

Figure 2.43 Topmost Help Window

Search for Help On...

Search for Help displays a scrolling list of topics. You select a topic by typing in a topic name (or part of a topic name). AFter you select and show a topic, click the **Go To** button to go directly to help for that topic.

Figure 2.44 Help for Specific Topics

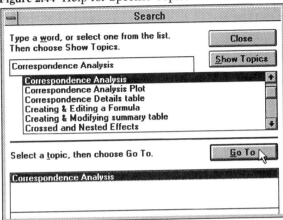

Statistical **G**uide

The **Statistical Guide** command displays a scrolling statistical index. When you click an analysis in the guide, directions on how to do the analysis appear as shown in **Figure 2.45**. The directions tell you the JMP menu, command and options to use for the analysis or topic you selected.

Figure 2.45 How to Perform Analyses

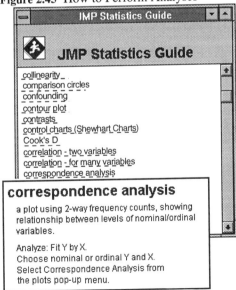

Help

How to Use Help

How to Use Help is the standard Windows help instructions found in most Windows applications.

About JMP

About JMP displays the JMP startup screen, which gives you license and version information.

See Chapter 7, "JMP Report Windows," for more information about using JMP help.

Chapter 3
JMP Data Tables

JMP data are organized in memory as rows and columns of a table referred to as the *data table*. The columns have names and the rows are numbered. An open data table is kept in memory, and you communicate with it through an active spreadsheet window. You can open as many data tables in a JMP session as memory allows. When you close a JMP data table, it is stored in a *file* on disk, sometimes called a *document*.

Macintosh Windows

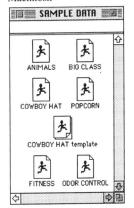
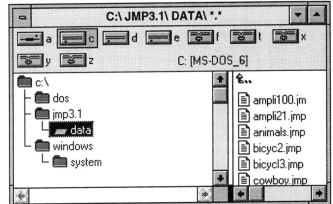

Commands in the **File**, **Edit**, **Tables**, **Rows**, and **Cols** menus give you a broad range of data handling operations and file management tasks such as flexible data entry, data validation, text editing, and extensive table manipulation. This chapter describes the elements of a JMP data table and its spreadsheet view, tells you how to create a new data table, describes characteristics of data, and explains how to use the spreadsheet effectively. Chapter 4, "The Tables Menu," covers each **Tables** menu command in detail.

Chapter 3
Contents

Elements of a JMP Data Table 71
   Cursor Forms 72
   Selecting and Deselecting Rows and Columns 73
Creating a New Table 75
   Adding and Deleting Rows 75
   Adding and Deleting Columns 76
Filling a Spreadsheet with Data 78
   Entering and Editing Data 78
   Importing Data 79
   Cutting and Pasting Data 82
   Data Validation 84
   Navigating with Keyboard Arrows 85
Exporting JMP Files 87
Characteristics of Data 88
   Data Types and Modeling Types 88
   Formats 89
   Dates 90
   Row States 91
Printing and Journaling Data 93
Using ClearAccess on the Macintosh 93

Elements of a JMP Data Table

JMP data are organized in memory as rows and columns of a table referred to as the *data table*. The data table opens in spreadsheet form as a standard window. The following is a list of some important data table features:

- Data can be manually keyed, calculated, imported from a text file, pasted into a JMP table, received as real–time external measurements (currently on the Macintosh only).
- Column names can have up to 31 characters and can use any keyboard character including spaces. The size and font for names and values is a preference setting you control.
- Character fields can be up to 255 characters long. You can move column boundaries and enlarge the column to view long values.
- Commands from the **Rows** and **Cols** menus add, move, hide, and delete rows and columns.
- There is no limit to the number of rows or columns in a data table. However, the table must fit in memory.

You can start JMP by opening an existing JMP data table, which displays the data table in spreadsheet form. To see the spreadsheet in **Figure 3.1**, open the TYPING DATA table (called typing.jmp under Windows) found in the SAMPLE DATA folder. The counts of rows and columns appear in the upper–left corner of the spreadsheet. A row number identifies each row, and each column has a column name.

This section describes the cursor forms that appear as you move the mouse on the spreadsheet, and tells how to select rows and columns, find rows, and use keyboard arrows. The **Characteristics of Data** section later in this chapter describes the role assignment and modeling type pop-up menus at the top of each column.

Figure 3.1 Active Areas of a JMP Spreadsheet View

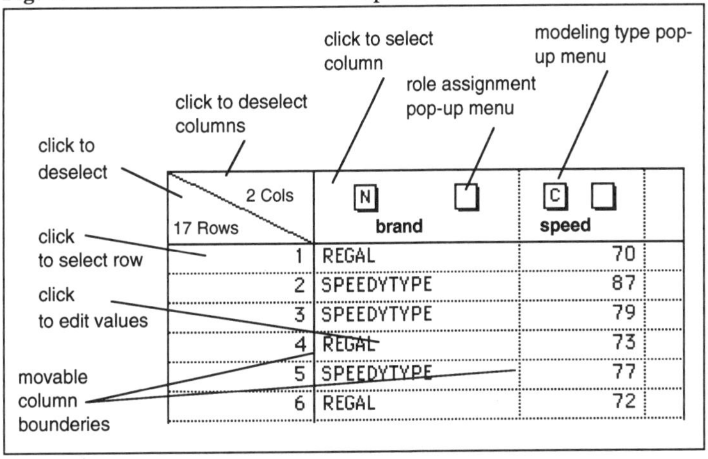

Cursor Forms

This section refers to a concept called *cell focus*. A table cell is focused when it is highlighted or contains the blinking vertical bar that indicates an insertion point. A focused cell responds to cell edit actions.

To navigate the spreadsheet, you need to understand how the cursor works. The cursor has different forms, and the actions it performs depend on its location in the spreadsheet:

Arrow Cursor

The cursor is the standard arrow when it is in the *modeling type box* or the *role assignment box* at the top of a column, in the *Cols* or *Rows* triangular areas in the upper–left corner of the spreadsheet, or in the empty portion of the table.

Cross Cursor

When the cursor is within a column heading or a data cell, it becomes a large cross indicating it is available to select text. When you click the cross cursor, that cell is focused and highlighted to show that its text is editable. The cursor then becomes an I–beam. Text in a locked column or a locked data table cannot be edited.

If you drag the cross cursor, it forms a stretch rectangle and selects the rows and columns formed by the cells within the rectangle.

Click twice on a column name to select it for editing.

I–beam Cursor

To edit text, position the I-beam within highlighted text. Click the I-beam to mark an insertion point, or drag it to select text for replacement. The I–beam deposits a blinking vertical bar to indicate a text insertion point or a highlighted area of text to be replaced. The default selection is the entire cell. Use the keyboard to make changes.

Open Cross Cursor

The cursor is a large open cross when you move it into a row or column selection area or a noneditable area of the spreadsheet such as a locked column cell. Click the open cross cursor to select a single row or column. Shift–click a beginning and ending range of rows or columns to select an entire array. Use COMMAND–click on the Macintosh (CONTROL–click under Windows) to make a discontinuous selection. See the next section, **Selecting Rows and Columns**, for more details.

Double Arrow Cursor

The cursor changes to a filled double arrow when on a column boundary. You can drag this cursor left or right to change the width of a column. However, this physical width does not affect the column field width specified in the **Column Info** dialog.

List Check and Range Check Cursors

The cursor changes form when you move the mouse over values in columns that have data validation in effect. It becomes a small, downward–pointing arrow on a column with list checking, and a large I–beam on a column with range checking. When you click, the value highlights and the cursor becomes the standard I-beam; you enter or edit data as usual. However, you are not allowed to enter invalid data values.

Selecting and Deselecting Rows and Columns

Commands from the **Rows** and **Columns** menus operate only on *selected* rows and columns. The selection of rows and columns is done by highlighting them. To highlight a row, click the space that contains the row number. To highlight a column, click the background area above the column name.

To extend the selection of rows or columns, drag across the array or SHIFT–click the first and last row or column range. Use COMMAND–click on the Macintosh (CONTROL–click under Windows) to make a discontiguous selection. To select rows and columns at the same time, drag the mouse across table cells to form a rectangular box. The table to the right in **Figure 3.2** shows the highlighted rows and columns that correspond to the selection rectangle on the left–hand table.

Figure 3.2 Select Rows and Columns by Dragging Across Cells

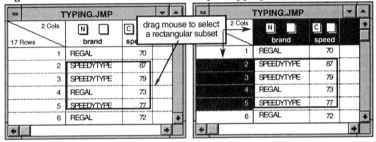

There are other ways to highlight specific rows:

- When you click on points in plots or bars of a graph, the corresponding rows highlight. The example in **Figure 3.3** shows a histogram with the SPEEDYTYPE bar highlighted, and the corresponding rows highlighted in the table. You can extend the selection of bars by SHIFT–clicking them. Use COMMAND–click on the Macintosh (CONTROL–click under Windows) to make a selection of discontiguous histogram bars, and use the brush tool or lasso tool to select multiple points in plots.

Figure 3.3 Select Rows and Columns Using a Histogram

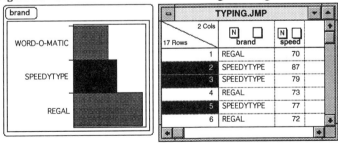

- The **Select** pop–up menu in the **Rows** menu has a **Where** option that lets you define the criteria for row selection. The specification for selection in the top dialog of **Figure 3.4** requests selection of all rows where the value of brand is equal to SPEEDYTYPE, which gives the same result as the histogram in **Figure 3.3**. The second dialog has the **Search selected rows only box** checked. The searches finds only the rows previously selected, and maintains the selection when speed is greater than 79. Using the **Search** command in this way is equivalent to the logical search request, 'select if brand is equal to SPEEDYTYPE and speed is greater than 79.'

Figure 3.4 Select Rows and Columns Using a Histogram

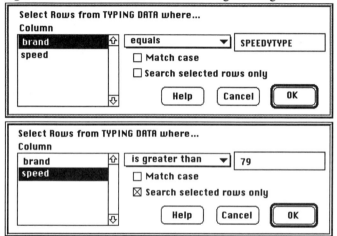

- You can use the **Group/Summary** command in the **Tables** menu to select subsets of rows. **Group/Summary** creates a JMP window that contains a *summary table*. This table summarizes columns from the active data table, called its *source table*. It has a single row for each level of a grouping variable you specify. Each row in the summary table identifies the corresponding subset of rows in the source table. When you highlight a summary table row, all corresponding rows highlight in its source table. See Chapter 4, "The Tables Menu," for more details about the **Group/Summary** command.

Locating Selected Rows

The **Locate Next** and **Locate Previous** commands in the **Rows** menu search for the next selected row below or above the *current row*. The current row is defined in any of the following ways:

- When you open a JMP data table, the first row is the current row.
- By default, the most recently edited row is the current row.
- To change the current row use OPTION–click on the Macintosh (CONTROL–click under Windows) anywhere within a row.
- A row found by **Locate Next** or **Locate Previous** is the current row.

If there are no selected rows in the spreadsheet, these two commands are dim.

Whenever you locate a selected row, the spreadsheet window scrolls to make that row visible, if necessary. The selection area of the row you locate also blinks briefly to distinguish it from any other selected rows in view.

You can use either of the locate commands repeatedly to step through the selected rows in the spreadsheet. If there are no more selected rows between the current row and the bottom of the spreadsheet, the **Locate Next** command alerts you with a beep. The location of the current row does not change. Likewise, if there are no more selected rows above the current row, the **Locate Previous** command alerts you.

Deselecting Rows and Columns

Use COMMAND–click on the Macintosh or CONTROL–click under Windows on the row or column selection area to deselect a rows or column. Or, click the triangular Rows or Cols area in the upper–left corner of the spreadsheet to deselect all rows or columns at once.

Note➡ **Invert Selection** is a **Rows** menu command that deselects all currently selected rows and selects all unselected rows. This menu item is useful when analyzing exclusive subsets of rows separately.

Creating a New Table

To process data with JMP, you must first create a JMP data table. The **New** command in in the **File** menu creates a new data table in memory and displays an empty spreadsheet with no rows and one numeric column, labeled Column 1, as shown to the right.

You fill it with values in one of the following ways:
- Create new rows and columns, and type data into the spreadsheet.
- Construct a formula to calculate column values.
- Import data from another application with the **Import** command in the **File** menu.
- Fill the table with values from the clipboard with the **Paste** or the **Paste at End** command in the **Edit** menu.
- Use a measuring instrument to read external measures (currently Macintosh only).

Adding and Deleting Rows

To add rows, use the **Add Rows** command from the **Rows** menu or double–click the triangular Rows area in the upper–left corner of the spreadsheet. This displays the Add Rows dialog shown in **Figure 3.5**. Alternatively, you can add blank rows at the end of the table by double–clicking anywhere in the empty table area beyond the last nonblank row. This fills the table with blank rows through the position of the cursor.

The Add Rows dialog prompts you to enter the number and location of rows you want to add. By default, new rows appear at the end of the table. To insert rows into a table, click either the **At Start** or the **After Row #** radio button.

If the values in an existing column are calculator computations, then the column's new cells automatically fill with values. Otherwise, the new cells have missing values.

Figure 3.5 The Add Rows Dialog

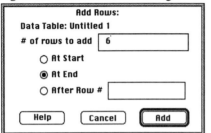 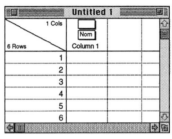

To delete rows from the spreadsheet, select the rows you want to delete and choose the **Delete Rows** command from the **Rows** menu. If you mistakenly delete rows, immediately select the **Undo Delete Rows** command from the **Edit** menu.

Adding and Deleting Columns

You can add columns to a JMP spreadsheet several ways:

- Use the **New Column** command in the **Cols** menu.
- Use the **Add Columns** command in the **Cols** menu.
- Add rows to an Attributes table.
- Using Save commands from a report window

Each of these ways to add columns is described in the following sections.

The New Column Command

To use the New Column dialog, select the **New Column** command from the **Cols** menu or double-click in the triangular Cols area in the upper left-corner of the spreadsheet. The dialog shown in **Figure 3.6** prompts you to name the new column and provide column characteristics. Optionally, you can click **Next** to define characteristics for additional new columns. When you click **OK**, the new columns appear in the data table.

The New Column dialog also gives you the option to compute the new column values with a formula. A calculator window for the new column displays whenever you select **Formula** from the **Data Source** pop-up menu and then click **OK**. You use this calculator to construct a formula. The formula can include existing columns, functions, parameters, and constants, and it can use conditional logic with comparison operators.

A column automatically locks and cannot be edited when its data source is **Formula**. To unlock a column, check the **Lock** box again. See Chapter 5, "Calculator Functions," and Chapter 6, "Using the Calculator," for details about the calculator.

Figure 3.6 The New Column Dialog

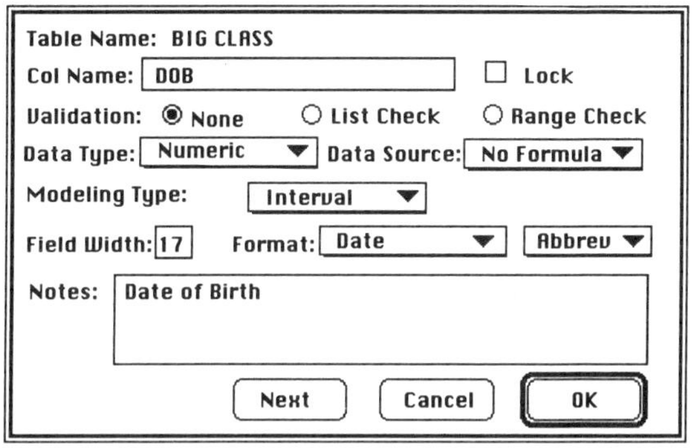

See Chapter 2, "The Menu Bar," for a description of all items in the New Column dialog. The **Characteristics of Data** section later in this chapter gives a description and examples of different kinds of data.

The Add Columns Command

The **Add Columns** command displays the dialog shown in **Figure 3.7**, which prompts you for the number of columns to add, and their location, field width, and type. Column characteristics are the same for all new inserted columns.

By default, the new column names are simply Column 1, Column 2, and so forth. If you enter text in the **Col Name Prefix** box, that text becomes the prefix of the new column names, as illustrated by the prefix name (**Item #**) used in **Figure 3.7**.

Figure 3.7 The Add Columns Dialog

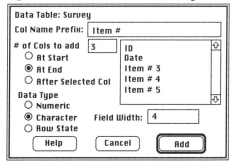

Using the **Add Columns** command to define multiple columns is different than using the **New Column** command because all columns you request with the Add Columns dialog have the same data characteristics. The New Column dialog creates each new column with the characteristics you enter.

The Attributes Table

You can use an *attributes table* to add new columns to a data table. The **Attributes** command in the **Tables** menu creates a special Attributes data table that contains editable information about the current data table, called its *source table*. The Attributes table has a row for each variable in its source table and a variable for each type of column characteristic. Editing the attributes table changes column characteristics in the Source table when you select the attributes table **Update** command.

You can add and delete rows from an Attributes table itself. When you add a row to an Attributes table, that row defines a new column in the Source table when you update it. Likewise, if you delete a row from an Attributes table and update its Source table, the corresponding Source table column is deleted.

There is further discussion of the **Attributes** command in Chapter 2, "The Menu Bar," and in Chapter 4, "The Tables Menu."

Report Window Save Commands

Report windows given by **Analyze** menu commands usually have save commands, which often add columns to the current data table. These commands are in the dollar ($) pop-up menu that accompanies each analysis window.

The Delete Columns Commands

To delete columns from the spreadsheet, select the columns you want to delete and choose **Delete Columns** from the **Cols** menu. If you mistakenly delete columns, immediately select the **Undo Delete Columns** command from the **Edit** menu.

Filling a Spreadsheet with Data

A JMP table is a flexible spreadsheet for preparing data. You can use the spreadsheet to enter data, edit cells, manipulate rows and columns, and perform other common spreadsheet tasks. This section tells you how to get data into a JMP table.

The description that follows distinguishes between *character* and *numeric* columns. Numeric columns must contain numbers only, with or without a decimal point. Character columns can contain any characters including numbers.

Entering and Editing Data

To enter data, move the cursor onto a cell. The cursor becomes a large cross. The cell highlights when you click in it, and the cursor becomes the I-beam. If you click again, the I-beam deposits a blinking text insertion bar. Typing replaces highlighted values or inserts values at the position of the insertion bar. The DELETE key on your keyboard deletes highlighted values or deletes values to the left of the insertion point.

Note➥ Text in a locked column or a locked data table cannot be edited.

Cells in Numeric Columns

On the Macintosh an empty numeric cell show a period. Under Windows the missing value indicator is a question mark. When you click anywhere in a numeric cell, it highlights and you can type in decimal numbers. Numeric values are right aligned. The **Format** pop–up menu in Column Info dialog lets you specify a number of decimals.

To edit existing numeric cells, click to highlight the cell and type to replace the text. To insert text or replace a portion of the existing value, click again to position the insertion point. This lets you insert new values or drag to highlight specific values.

Cells in Character Columns

To enter values into an empty character column cell, click to position the insertion bar and type data values. Values in character cells are left aligned. To edit existing values, highlight the values you want to replace or click to position the insertion bar. Then use the keyboard to make changes.

Finding and Replacing Cell Values

The **Search** command in the **Edit** menu displays the submenu shown to the right. Search and replace actions deal only with character strings. Numbers are treated as text, and appear to the **Search** command as they show in the data table. For more information, see **The Edit Menu** in Chapter 2, "The Menu Bar."

Find...	⌘F		Find	Ctrl+F
Find Next	⌘G		Find Next	Ctrl+G
Replace	⌘R		Replace	Ctrl+R
Replace & Find Next	⌘H		Replace & Find Next	Ctrl+H
Replace All			Replace All	

Importing Data

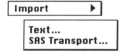

The **Import** command in the **File** menu has a pop–up menu that lets you read (import) data from a standard text file or a SAS transport file into a JMP table.

If you a Macintosh with the ClearAccess query tool installed, the Import pop–up menu shows the additional ClearAccess selections, described in the last section of this chapter.

Text–formatted Files

JMP automatically creates a data table from a text file of values arranged in rows and columns. The **Text** selection on the Import dialog displays a modified file selection dialog as shown in **Figure 3.8**. To import a text file, you need to specify the end–of–field and the end–of–line delimiters by clicking appropriate check boxes.

Figure 3.8 The Text Import Dialog

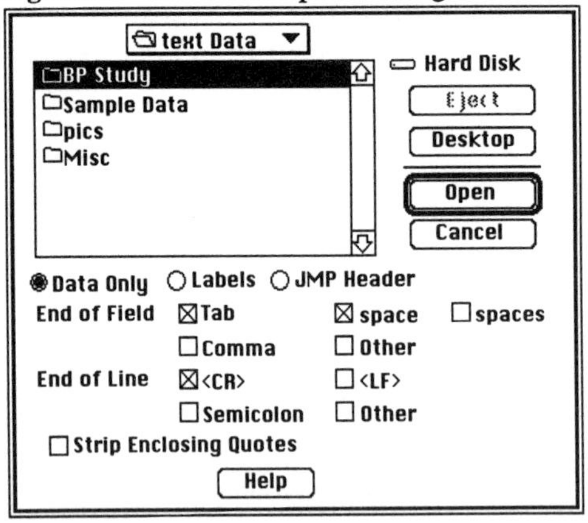

Macintosh Text Import Dialog

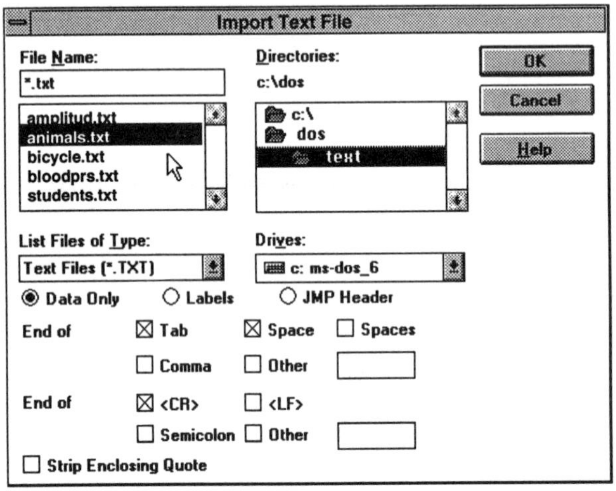

Windows Text Import Dialog

Radio buttons on the dialog give JMP specific information about the text file:

- ◉ **Labels**
  tells JMP the incoming data are from a text–format file that contains column labels as the first record of the data. These labels become column names in the new JMP data table. For example, the file shown in **Table 3.1** illustrates a text file with column labels as the first record in the file.

Table 3.1 Example of Text File with Labels

```
name     sex age height weight
ALFRED    M   14  69.0  112.5
ALICE     F   13  56.5   84.0
                  .
                  .
WILLIAM   M   15  66.5  112.0
```

When you select the **Labels** radio button and open the text file above, JMP names the columns of the data table with fields (up to 31 characters) from the first line of the text file. The **Import** command creates the JMP data table shown in **Figure 3.9**. The new data table has the same name as its original text file.

Figure 3.9 Label Option for Text Files with Column Labels

	CLASS			
	[N] Name	[O] sex	[N] age	[C] height	[C] weight
1	ALFRED	M	14	69	112.5
2	ALICE	F	13	56.5	84
3	BARBARA	F	13	65.3	98
4	CAROL	F	14	62.8	102.5

◉ **JMP Header**
indicates that the incoming data are from a text–format file that contains column–header information as exported from a JMP file.

The check boxes identify the end–of–field and end–of–line delimiters:

End of Field
designates the character that identifies the end of each field in the incoming file. This delimiter can be a tab, one or more blank spaces, any character, or any combination of the above. When you choose **Other**, an editable text box appears where you can enter a character or its hexadecimal representation.

End of Line
selects the character or key that identifies the end of each line. You can use the RETURN key, the line feed character, a semicolon, or another character of your choice. The **Other** option works the same as it does with **End of Field** described above.

Note➧ To enter a hexadecimal value, precede the value with '0x'.

SAS Transport Files

The **SAS Transport** selection in the **Import** pop–up menu creates a data table from a file that is in the SAS transport format. A SAS transport file is created whenever you

- use the SAS System to write an existing SAS data set in transport format
- use the **Save As** command in the **File** menu and click the **SAS Transport Format** radio button before exporting the JMP table.

The **SAS Transport** option on the Import dialog displays the dialog shown in **Figure 3.10** for creating a JMP data table from a member of a SAS library. If you click the **Open All Members** box, JMP creates a data table for each member in the SAS data library.

Figure 3.10 The SAS Transport Import Dialog

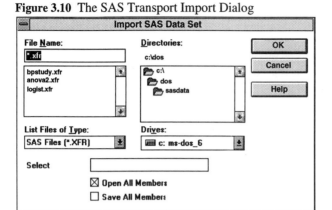

Windows
SAS Transport Import
Dialog

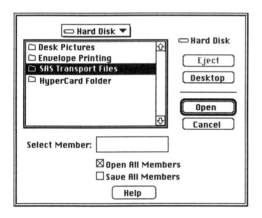

Macintosh
SAS Transport Import Dialog

Cutting and Pasting Data

Edit menu commands have different effects depending on row, columns, and cell selection in the current data table. These commands either affect selected rows and columns or affect a single focused cell:

- If there is a focused cell, all **Edit** commands apply only to that cell.
- The **Edit** commands affect all values in selected rows if no columns are selected. They affect all values in selected columns (except the header field) if no rows are selected.
- If you select both rows and columns, **Edit** commands affect the subset of cells defined by the intersection of those rows and columns.
- If there is no focused cell or selected region, all **Edit** commands are dim except the **Paste at End** command.

The **Edit** menu commands function as follows:

Copy
copies selected fields (either data values or a column heading) from the active data table to the clipboard.

Copy as Text
copies only text data from a report on an Analyze or Graph platform.

Cut
copies selected fields (either data values or a column heading) from the active data table to the clipboard and replaces them with missing values. **Cut** is equivalent to **Copy**, then **Clear**.

Clear
replaces selected fields (either data or a column heading) in the active data table with missing values but does not copy them to the clipboard.

Paste
copies data from the clipboard into the selected area of a JMP data table. If the selected area is larger than the contents of the clipboard, JMP fills the selected cells by cycling the clipboard contents as often as necessary. If the selected area is smaller, JMP pastes as much of the clipboard contents as possible.

Undo
cancels the effect of the most recent reversible command. If **Undo** is available, it displays in the **Edit** menu as **Undo** *command*, where *command* is the reversible command. Most destructive spreadsheet operations (such as cutting, pasting, and deleting) are reversible. However, **Undo** is dim when it is unavailable. After you select **Undo**, it changes to **Redo** *command.*

Paste at End
pastes data from the clipboard at the end of selected rows or selected columns (but not both) in the active spreadsheet. If both rows and columns are selected, the **Paste at End** command is dimmed. The **Paste at End** command reads only data in text format from the clipboard. However, it can paste data into character, numeric, or row state columns as long as these corresponding columns have matching data types. No other clipboard type formats are supported.

The **Paste at End** Command copies character strings from the clipboard, pasting each string that is separated by a tab character into a separate spreadsheet cell. Likewise, each return character in the clipboard represents the end of a spreadsheet row. Any cells copied from JMP are automatically stored in this format.

If one or more rows are selected when you choose **Paste at End**, the command adds columns to the right of existing columns. The command pastes the number of rows from the clipboard that corresponds to the number of selected rows in the spreadsheet. All other rows in the clipboard are ignored, and the cells corresponding to other rows in new columns fill with missing values.

If one or more columns are selected when you choose **Paste at End**, the command adds rows to the bottom of existing rows. The command pastes the number of columns from the clipboard that corresponds to the number of selected columns in the spreadsheet. All other columns in the clipboard are ignored, and the cells corresponding to other columns in new rows fill with missing values.

Paste at End gives you an error message if you select rows in a spreadsheet that include row state columns. You also see an error message if you **Paste at End** to numeric columns when the clipboard contains nonnumeric character data.

If there are no selected rows or columns, **Paste at End** behaves as if all existing columns are selected. Therefore, you can use this command to import data because it creates the rows and columns needed for the data in the clipboard.

The OPTION or ALT key

Note → You can copy selected information from an **Analyze** or **Graph** report window into a JMP table as follows:

- Select a table in a report window, hold down the OPTION (or ALT) key and choose **Copy as Text**.

- Open a new data table Then hold down the OPTION (or ALT) key and choose **Paste at End**. The values in the first line of information on the clipboard become column headers. The remaining information becomes the data table content.

As an example, the table to the left in **Figure 3.11** is the result of a JMP analysis. The scissors tool selects the table and OPTION (or ALT)–**Copy as Text** copies the table text and column headers (Term and Estimate) to the clipboard. Then, use OPTION (or ALT)–**Paste at End** when a new data table is active to see the table shown to the right.

Figure 3.11 Parameter Estimates Table Copied to a JMP Data Table

Parameter Estimates	
Term	Estimate
Intercept	66.5
HBARS[down-up]	.5
DYNAMO[off-on]	-6
SEATS[down-up]	1.75
TIRES[hard-soft]	-1.25
GEAR[low-medium]	-11.25
RAINCOAT[off-on]	0.25
BRKFAST[no-yes]	0.5

Untitled 1 — 2 Cols, 8 Rows

	Term	Estimate
1	Intercept	66.5
2	HBARS[down-up]	0.5
3	DYNAMO[off-on]	-6
4	SEAT[down-up]	1.75
5	TIRES[hard-soft]	-1.25
6	GEAR[low-medium]	-11.25
7	RAINCOAT[off-on]	0.25
8	BRKFAST[no-yes]	0.5

Data Validation

The **Validation** radio buttons in the New Column dialog (see **Figure 3.6**) let you set acceptable values or an acceptable range of values for a column. The **List Check** radio button displays the dialog shown to the left in **Figure 3.12** for you to enter a list of valid values. If the column type is numeric, you can click the **Range Check** radio button and complete the dialog on the right to specify a range of values and range limit conditions. For a single-sided range check, omit one of the bounds (upper or lower).

Figure 3.12 Data Entry Validation Dialogs

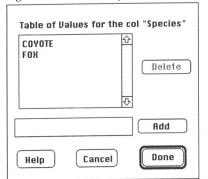 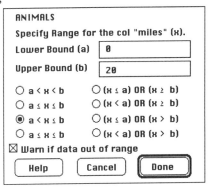

↕ The cursor changes to the list check form when positioned over any editable area (cell or column name) in a column with data validation checking in effect. When you click a cell, a blinking vertical bar appears for entering or editing text as usual. However, you can only enter values listed on the validation list. A beep and a message warn if you try to enter invalid text. If you COMMAND–click a cell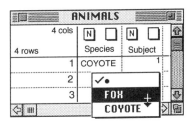
(CONTROL–click under windows), a pop–up menu of acceptable values shows. This lets you select the cell value from this menu instead of typing it into the cell. If you select a value from the menu when the list check cursor is over a column name, that value enters into all cells in the column.

⫯ The cursor changes to a large I-beam on numeric column values with a specified range and range condition. Optionally, a beep warns if you try to enter invalid text.

Navigating with Keyboard Arrows ← ↑ → ↓

If you prefer to use the keyboard instead of a mouse, you can navigate efficiently about an active spreadsheet with keystrokes. Most keyboards have four arrow keys on the lower–right half of the keyboard. Pressing one of these keys or the TAB, RETURN, or ENTER key while a spreadsheet cell is focused, moves the cursor and focuses a new cell. Holding down the SHIFT and OPTION keys on the Macintosh with arrows modifies their behavior. See **Table 3.2** for a summary of keyboard arrow actions.

Table 3.2 Keyboard Commands and Actions

Keystroke	Moves Focus or I–beam
→	one highlighted cell to the right, wraps through lower–right cell, or insertion point moves one character to the right through end of a cell value
←	one highlighted cell to the left, wraps through leftmost column name, or insertion point moves one character to the left to beginning of value
↑	highlights one cell above, up through same column's name
↓	highlights one cell down, to last cell in the same column
SHIFT – →	highlights cell characters one at a time to the right of the insertion point
SHIFT – ←	highlights cell characters one at a time to the left of the insertion point
SHIFT – ↑	toggles highlighting the focused cell (locates cell focus)
SHIFT – ↓	toggles highlighting the focused cell (locates cell focus)
OPTION – →	highlights rightmost visible cell, or moves insertion point one character to the right (not available under Windows)
OPTION – ←	highlights leftmost visible cell, or moves insertion point one character to the left (not available under Windows)
OPTION – ↑	highlights first cell in column (not available under Windows)
OPTION – ↓	highlights last cell in column (not available under Windows)
TAB	highlights adjacent cell right, wraps down and creates new row at end
SHIFT–TAB	highlights adjacent cell left, wraps upward through column names
RETURN	same as down arrow
SHIFT–RETURN	same as up arrow
ENTER	moves cell focus down and highlights cell
SHIFT–ENTER	moves cell focus up and highlights cell, through the column name

Exporting JMP Files

To export a JMP file, use the **Save As** command in the **File** menu. The Save As dialog. has several export options. You can write files in standard JMP file format, in standard text format for use in other applications, or as SAS transport files that can be read automatically by the SAS System. **Save As** saves the current data table to a file after prompting you for a name and disk location.

Figure 3.13 Save As Dialog and Text Formatting Options

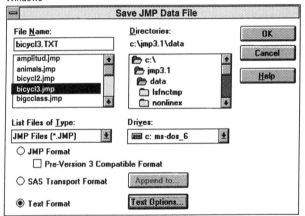
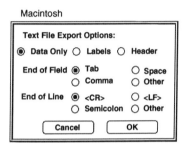

The **Text Format** option writes the rows and columns of a JMP data table to a rectangular text file using the text formatting options you specify. However, the **Labels** option creates a special file format that saves the data table column names. For example, you can create a text file from the data table shown in **Figure 3.9**. The **Labels** option copies the column names as fields in the first line, followed by data lines.

The resulting text file is illustrated in **Table 3.3**. Tab characters (represented by right arrows) separate the fields, and carriage returns (represented by ¶) delimit the lines. The leading blanks before the numeric values show as periods. A missing numeric value shows as a bullet (•), and missing character values show as blanks.

Table 3.3 Illustration of Text File Created with Output Labels Option

name➡	sex➡	age➡	height➡	weight¶
ALFRED➡	M➡	.....14➡	......69➡	...112.5¶
ALICE➡	F➡	.....13➡	........•➡	.......84¶

and so forth

Characteristics of Data

The columns of a JMP table can contain different kinds of information. However, all information in a single column must be of a similar type. You create a new column with the **New Column** command, and define its characteristics with the New Column dialog.

JMP correctly analyzes data according to column characteristics. This section discusses the following attributes of data table columns:

- data types and modeling types
- data storage and formats
- a special format for date values.

See **The Cols Menu** section in Chapter 2, "The Menu Bar," for illustrations and discussion of **The New Column** command and the Column Info dialog.

Data Types and Modeling Types

The Column Info dialog lets you give JMP a description of each column's data characteristics by assigning the column one of three *data types* and one of three *modeling types*:

Data Type

The data type of a column determines how its values are formatted in the spreadsheet, how they are stored internally, and whether they can be used in calculations. The three data types are called *numeric*, *character*, and *row state*.

Modeling Type

The modeling type assigned to a column tells JMP how to treat its values in analyses. The three modeling types are called *continuous*, *ordinal*, and *nominal*. The modeling type of a column determines how it is treated by the **Analyze** menu commands.

Column Characteristic Combinations			
	Data Type		
Modeling Type	numeric	character	row state
continuous	YES	NO	N/A
ordinal	YES	YES	N/A
nominal	YES	YES	N/A

The table above illustrates possible combinations of data types and modeling types. These combinations are described next.

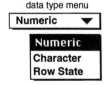
data type menu

The first characteristic assigned to a column is its data type. All columns have one of the three data types listed by the **Data Type** pop-up menu in the **Column Info** dialog. You can change a column's data type as long as the column values are as follows:

- **Numeric** columns must only contain numbers, with or without a decimal point. A numeric column can have any modeling type.

Numeric column values display right aligned and can be formatted with the **Format** pop–up menu. You can change a numeric column to character or row state with the **Data Type** pop–up menu.

- **Character** columns can contain any characters including numbers. In character columns, numbers are seen as characters only and are treated as discrete values instead of as continuous values. You can change a character column to numeric. Any values that are numbers convert correctly, and character values become missing.
- **Row State** columns contain special information that can affect the appearance of graphical displays. A row state column does not have a modeling type because its values are not used in analyses. See the **Row States** section later in this chapter for details about using row state information.

You also assign both numeric and character columns one of the three modeling types listed by the **Modeling Type** pop–up menu. This menu is in the Column Info dialog and at the top of each spreadsheet column. JMP uses the column's modeling type to determine how to analyze its values.

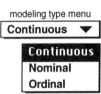

You can change a column's modeling type and see the data treated in a different way as long as the assigned modeling type corresponds to the data types described previously. The following list describes modeling types:

- **Continuous** columns must contain numeric values and be assigned the numeric data type. Continuous values are treated as continuous measurement values. JMP uses the numeric values directly in computations.
- **Ordinal** columns can have either numeric or character data types. JMP analyses treat ordinal values as discrete categorical values that have an order. If the values are numbers, the order is the numeric magnitude. If the values are character, the order is the sorting sequence.
- **Nominal** columns can have either numeric or character data types. All values are treated in JMP analyses as though they are discrete values with no implicit order.

Formats

Numeric column values are formatted by the **Format** pop–up menu selections found in the Column Info dialog:

- **Fixed Decimal** displays all column values rounded to the number of decimal places you specify.
- **Best** format means JMP considers the precision of each column value and chooses the best way to show it.
- **Date & Time** formats are described in the next section.

Note➔ JMP uses the decimal point defined by your localized system software. For United States systems, this is the period character.

Dates & Times

When you assign the **Date** format to a numeric column, JMP assumes its values are the *number of seconds since January 1, 1904*. For example, there are 1,234,567,890 seconds between January 1, 1904 and February 13, 1943.

You can leave the date values displayed as (very large) numbers, but they would not convey much date information. When you choose the **Date & Time** format, an additional pop-up menu appears and lists specific date and time representations:

- The **Short** date format displays a JMP date as *dd/mm/yy* (2/13/43).
- The **Long** date format displays a JMP date value as *weekday, month day, year* (Saturday, February 13, 1943).
- The **Abbrev** date format display is the same as the **long** format except that the weekday and month have three-character abbreviations (Sat, Feb 13, 1943).
- The **Date:HH:MM** and **Date:HH:MM:SS** format display a date value as a **Short** date followed by the number of hours, minutes, and seconds after midnight of that date.
  The formatted values for this example are 2/13/43 11:31 PM and 2/13/43 11:31:30 PM.
- The **:Days:HH:MM** and **:Days:HH:MM:SS** formats show the number of days, hours, Minutes, and seconds since January 1, 1904. The formatted results for December 31, 1995 are :14288:23:31 and :14288:23:31:30.

date and time formats

Date & Time ▼
✓ Short
Long
Abbrev
Date:HH:MM
Date:HH:MM:SS
:days:hrs:mins
:days:hrs:mins:secs

You may need to increase the column width to see the entire formatted value for **Long** and **Abbrev** date formats, and for date-time formats.

Note→ The month and day names, the date field separators, and the order of date elements depend on your localized system software, which uses the standards in your country. Also, if you are using the Macintosh operating System 7.1 or later, JMP uses the date forms you define in the Date and Time control panel for Long, Short and Abbreviated.

Selecting a date or time format for a column also lets you enter values using any representation recognized by your machine. When you press RETURN, JMP stores the date entry internally as the numeric number of seconds between January 1, 1904 and the date you entered. The date displays according to the format you assigned to the column.

The JMP calculator offers full support for dates with functions that accept date columns as arguments and return date-related elements such as day of the week, day of the year, week of the year, or month. See Chapter 5, "Calculator Functions," for a description of date functions.

If you are importing variables from a SAS transport file, JMP looks for a SAS date format and translates it to a JMP date column. When you are exporting data to a SAS file, JMP date columns become SAS date values with the appropriate SAS format.

Row States

Row states are characteristics associated with a row. They can distinguish subsets of your data, exclude data from analyses, and customize the appearance of graphical displays. Row state commands are in the **Rows** menu and affect only highlighted rows. Row state assignments appear in the row number area at the left of the spreadsheet (see **Figure 3.13**).

Active row state characteristics can be saved permanently with the data table in a special row state column. To save row state information, create a new column and assign it the **Row State** data type.

There are two ways to fill a column with row state information:
- Use the calculator to assign row state information to the column based on a formula.
- Select the **Copy from Row State** command in the ➡ pop–up menu at the top of the row state column (see **Figure 3.13**). This copies the active row state assignments from the row number area to the column.

JMP uses the following row states:

Exclude/Include ⊘
is a toggle that excludes selected rows from statistical analyses. To do this, select the rows that contain unwanted values and choose **Exclude/Include** from the **Rows** menu. Data remain excluded until you choose **Exclude/Include** again for those selected rows.

Warning➡ Excluded data are not automatically hidden in plots even though they are excluded from calculations in text reports and graphical displays.

Hide/Unhide 👓
is a toggle that suppresses the display of points in all scatter plots. To hide data, select rows in the spreadsheet and choose **Hide/Unhide** from the **Rows** menu. For example, you can exclude points from analysis and then hide those same points in scatterplots. The data remain hidden until you choose **Hide/Unhide** again for selected hidden rows.

Warning➡ Hidden points are not automatically excluded from statistical computations that affect text reports and graphical displays even though they are not displayed in the plots. To exclude hidden observations from analyses you must also use assign them the **Exclude/Include** row state.

Label/Unlabel
is a toggle that labels points on all scatterplots. To label points, select the rows containing the points and use the **Label/Unlabel** command. This row state remains in effect until you choose **Label/Unlabel** again for those selected rows.

JMP uses the row number as the label value on scatterplots if you don't assign a label column. However, you can use the values of any column to be labels that identify points in plots by assigning the column the Label role.

Colors
lets you assign any colors to highlighted rows (only visible on color monitors). The points in scatterplots and spinning plots display using the color you select from the

Colors palette. The active color assigned to a row displays next to the row number in the spreadsheet.

Markers ✖

assigns a character to replace the default dot in scatterplots and spinning plots. There are eight markers available in the **Markers** palette. Each active marker displays next to its row number in the spreadsheet.

Selection Status

is a row state characteristic that can be saved in a row state column. However, selection status is not assigned using the **Rows** menu. For details about how to select rows, see **Selecting Rows and Columns** earlier in this chapter. In the spreadsheet, the row number area of a selected row is highlighted.

The **Select** command in the **Rows** menu is useful for selecting all rows or a subset of rows with specific row state characteristics. **Select All Rows** selects all the rows in a JMP data table.

You can select rows with the excluded, hidden, or labeled row state characteristic using the **Select** submenu:

- **Select Excluded** selects all excluded rows regardless of their current selection status and deselects any other previously selected rows.

- **Select Hidden** selects all hidden rows regardless of their current selection status and deselects any other previously selected rows.

- **Select Labeled** selects all labeled rows regardless of their current selection status and deselects any other previously selected rows.

To activate a row state column, select either the **Copy to Row State** or **Add to Row State** command in the ← pop–up menu at the top of the row state column. The active row state icons appear in the row number space as shown in **Figure 3.14**.

Figure 3.14 Row State Column Buttons

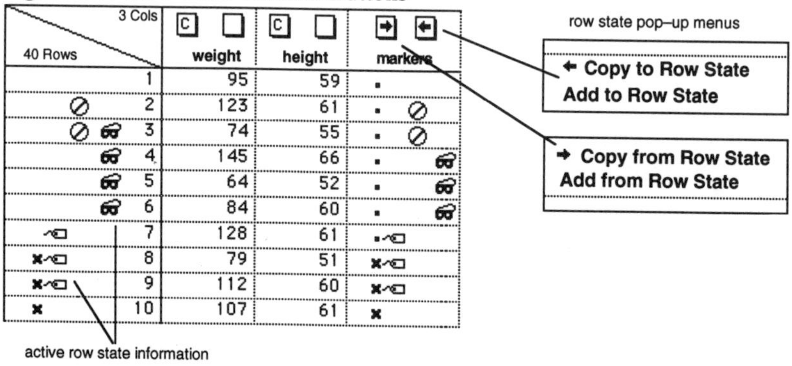

Copy to Rowstate

copies the row state characteristics from the row state column to the row number area, which makes them active. **Copy to Rowstate** replaces existing active row states.

Add to Rowstate
adds the row state characteristics from the row state column to the row number area, which makes them. **Add to Rowstate** preserves existing active row states.

Copy from Rowstate
copies the active row state characteristics from the row number area to the row state column. **Copy from Rowstate** replaces existing row state values in the column.

Add from Rowstate
adds the active row state characteristics from the row number area to the row state column. **Add from Rowstate** preserves existing row state values in the column.

Printing and Journaling Data

The **Journal** command from the **File** menu lets you save the spreadsheet in a journal file. You can journal the entire spreadsheet or a subset of data defined by selected rows and columns. See Chapter 7, "Report Windows," for information about journaling.

Note→ A column containing row state data cannot be saved in a journal file.

Using ClearAccess on the Macintosh

If you have the ClearAccess query tool installed on your Macintosh, ClearAccess selections show on the **Import** command submenu.

Figure 3.15 Import Command with the ClearAccess Selection

JMP Version 3 includes special support for the ClearAccess query tool from ClearAccess Corporation. The ClearAccess package allows you to create, edit, and submit database queries to a variety of popular industry-standard relational databases. Usually, the ClearAccess tool returns tables of data downloaded from a database to the Macintosh Clipboard or to a text file. However, JMP allows direct importing of data from a remote database through the ClearAccess application. JMP executes ClearAccess scripts and the resulting data appears in a new JMP data table without any intermediate steps.

The ClearAccess selections are briefly described here. See the documentation shipped with your ClearAccess package for details about writing, editing, and executing ClearAccess scripts.

Run ClearAccess Script

prompts the user for the name of an existing ClearAccess script and asks the ClearAccess application to execute it. If the ClearAccess application is not running, it is launched and brought to the foreground for the duration of the query. During the query, you will occasionally see status messages displayed in the ClearAccess application's window. When the script finishes execution, any data extracted from the remote database is placed into a new JMP data table named **ClearAccess Import**. You can rename the table, and use it as you would any other JMP table.

Edit ClearAccess Script

brings the ClearAccess application to the foreground and displays an editing window. You can then edit your script and save the changes. Details on writing ClearAccess scripts can be found in the ClearAccess documentation.

Launch ClearAccess

This option in the Import menu is a shortcut for users who need to use the ClearAccess application, but do not want to execute or edit a script immediately. When choosing this option, the ClearAccess application is launched and is brought to the foreground.

Chapter 4
Manipulating Tables: The Tables Menu

The **Tables** menu has commands that perform a wide variety of data management tasks on JMP data tables. These commands let you sort, subset, stack, or split table columns, join two tables side by side, concatenate multiple tables end to end, and transpose tables. You can also create Summary tables for group processing and summary statistics and an attributes table for easy table maintenance.

The **Design Experiment** command creates a table of runs for the experimental design you specify. **Design Experiment** accesses JMP's design–of–experiments module. See Chapters 25 through 30 in the *JMP Statistics and Graphics Guide* for complete documentation of Design of Experiments in JMP.

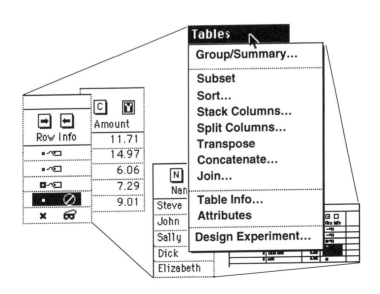

Chapter 4
Contents

The Tables Menu	97
The Group/Summary Command	98
The Summary Table	98
Subset Analysis	101
Summary Statistics	102
The Subset Command	106
The Sort Command	107
The Stack Columns Command	108
The Split Columns Command	110
The Transpose Command	111
A Simple Transpose	112
Transpose with a Label	112
Transpose Groups	113
The Concatenate Command	114
Concatenate Tables with the Same Column Names	114
Concatenate Tables with Different Column Names	115
The Join Command	117
Join by Row Number	117
Keep a Subset of Columns	118
Join by Matching Columns	120
Cartesian Join	123
The Table Info Command	125
The attributes Command	125
Attributes Table Columns	126
Adding, Deleting, and Moving Source Table Columns	128
Updating a Source Table	128
The Design Experiment Command	129

The Tables Menu

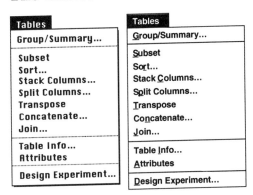

The commands in the **Tables** menu modify JMP data tables or create a new table by combining one or more tables:

Group/Summary
creates a JMP window that contains a *summary table*. This table summarizes columns from the active data table, called its *source table*. The summary table has a single row for each level of a grouping variable you specify. Optionally, you can add columns of summary statistics to this table.

Subset
creates a new data table that is a subset of selected rows and columns from the active spreadsheet.

Sort
sorts a JMP data table by one or more columns in either ascending or descending order.

creates a new data table by combining specified columns from the active data table into a single new column.

Split Columns
creates a new data table by dividing a specified column into several new columns according to the values of an ID variable.

Transpose
creates a new data table that is the transpose of the active data table. The columns of the active table are the rows of the new table, and its rows are the new table's columns.

Concatenate
creates a new data table from two or more data tables by combining them end to end.

Join
creates a new data table by joining (merging) two tables side by side.

Table Info
displays a dialog to enter notes or attributes for the current data table.

Attributes
creates a special table that contains editable information about the current data table. Editing characteristics in the attributes table can change them in the source table.

Design Experiment
creates a table of runs for the experimental design you specify. See Chapter 25 through Chapter 30 in the *JMP Statistics and Graphics Guide* for complete documentation of this command.

The Group/Summary Command

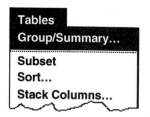

The **Group/Summary** command creates a JMP window that contains a *summary table*. This table summarizes columns from the active data table, called its *source table*. It has a single row for each level of a grouping variable you specify. When there are several grouping variables, the summary table has a row for each combination of levels of all variables. Each row in the summary table identifies its corresponding subset of rows in the source table.

You can expand the summary table by adding columns of summary statistics for any numeric column in the source table. Initially, a summary table contains the same information as frequency reports for nominal and ordinal variables given by **Analyze** menu commands. However, you can use the summary table produced by the **Group/Summary** command for several purposes:

Create a table of summary statistics
The **Stats** pop-up menu on the Group/Summary dialog lists standard univariate descriptive statistics. You can add columns of descriptive statistics to the summary table for any numeric column in the source table.

Analyze subsets of data
When you highlight rows in the summary table, the corresponding rows highlight in its source table. If the summary table is in By mode, the highlighted rows in the source table identify subsets. Commands on the **Analyze** menu recognize subsets identified by a summary table. A single **Analyze** menu command produces a separate report for each subset selected by the summary table.

Plot and chart data
The **Bar/Pie Charts** selection in the **Graph** menu lets you display summarized data easily. However, **Bar/Pie Charts** requires unique values of any column given an X role. You may need to preprocess raw data tables with the **Group/Summary** command and produce summary tables to use the **Bar/Pie Charts** effectively.

The Summary Table

The example data used to illustrate the **Group/Summary** command is the JMP table called companys.jmp in the data folder under Windows (COMPANIES in the SAMPLE DATA folder on the Macintosh) shown in **Figure 4.1**. It is a collection of financial information for 32 companies. The first column (Type) identifies the type of company with values "Computer" or "Pharmeceut." The second column (Size Co) categorizes each company by size with values small, medium, and big. These two columns are typical examples of grouping information.

Figure 4.1 JMP Table for Grouping Examples

7 Cols / 32 Cols	Type	Size Co	Sales($M)	Profits($M)	# Employ	profit/emp	Assets
1	Computer	small	855.1	31.0	7523	4120.70	615.2
2	Pharmaceut	big	5453.5	859.8	40929	21007.11	4851.6
3	Computer	small	2153.7	153.0	8200	18658.54	2233.7
4	Pharmaceut	big	6747.0	1102.2	50818	21690.02	5681.5
5	Computer	small	5284.0	454.0	12068	37620.15	2743.9
6	Pharmaceut	big	9422.0	747.0	54100	13807.76	8497.0

The **Group/Summary** command displays the dialog in **Figure 4.2**. To create the summary table shown in **Figure 4.3**, first select the variable Type in the **Columns** list of the dialog. Click **Group** to see it in the grouping variables list to the right. You can select as many grouping variables as you want. Click **Done** to see the summary table.

The **Tables Menu** section in Chapter 2, "The Menu Bar," describes each element of the Group/Summary dialog.

Figure 4.2 The Group/Summary Dialog

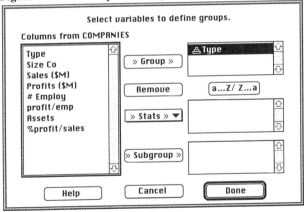

The new summary table appears in an active window. This table is linked to its source table. It is not saved when you close it unless you use the **Save As** command to give it a name and location.

Figure 4.3 Grouping Dialog and Summary Table

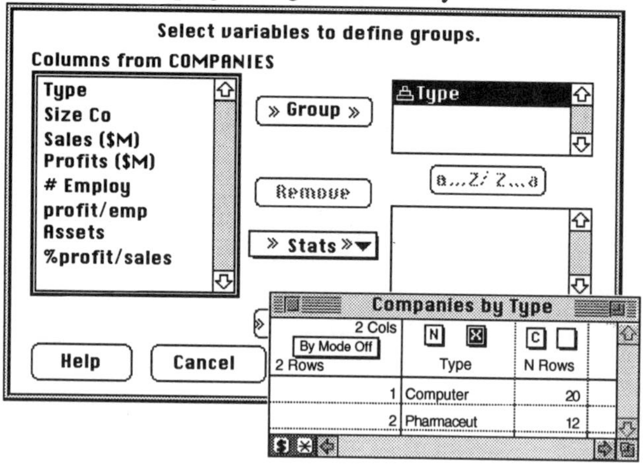

A summary table has one column for each grouping variable plus a column named N with counts for each grouping level. The example above shows 20 computer companies and 12 pharmaceutical companies. The values of the grouping variables are listed in either ascending or descending sort sequence. To specify the order of grouping levels, select the appropriate column in the grouping variable list and clicking the **a...Z/Z...a** toggle. By default, the grouping levels appear in ascending order.

The example in **Figure 4.4** shows the COMPANIES data table summarized by both type and size of company, with the values of size listed in descending order within company type. To follow along with the example, use the COMPANIES (or COMPANYS.JMP) data table to create three summary tables grouped by Type, by Size Co, and by both of these columns.

The summary table behaves like a graphical display window in the following ways:

- When you highlight rows in the summary table, the corresponding rows highlight in its source table.
- There is a dollar ($) option pop–up menu at the lower left of the window. This option accesses the Group/Summary dialog so that you can add statistical summary columns to the table at any time.
- The summary table, like a graphical display, is not saved when you close it. However, you can use the **Save As** command to specify a name and disk location for the table.

Figure 4.4 Second Grouping Variable in Descending Sequence

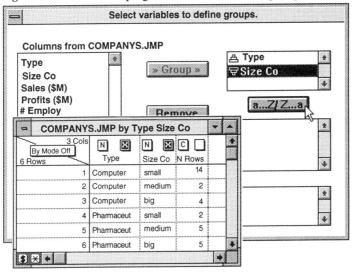

The summary table has two modes, accessed by the **By Mode** pop–up menu in the upper–left corner of the table:

- When By-Mode is on, **Analyze** menu commands recognize highlighted subsets in the source table identified by selected of rows in its summary table. **Analyze** menu commands produce a separate report window for each source table subset selected in the summary table.
- When By-Mode is off, **Analyze** and **Graph** menu commands and the **Transpose** command, discussed later in this chapter, act on the summary table itself.

Subset Analysis

When By-Mode is off, you can use **Tables** or **Analyze** menu commands on the summary table itself. The example shown in **Figure 4.5** uses the **Distribution of Y** command in the **Analyze** menu to show a histogram of Size Co. The N column in the summary table is automatically given the Freq role when JMP creates the summary table. This result is the same as a **Distribution of Y** on Size Co in the source table.

Figure 4.5 Distribution of Y on Summary Table with No Rows Selected

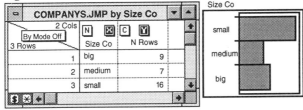

When you highlight rows in the summary table, all corresponding rows highlight in its source table. When By-Mode is on, **Analyze** and **Graph** menu commands then recognize subsets of these highlighted rows, as defined by the highlighted rows in the summary table. For example, the highlighted rows for the small and big company sizes in the summary table (**Figure 4.6**) identify the two subsets in the source table with those values of Size Co. Highlighting the same rows in the source table itself does not create any subsetting information.

If all the rows of a group have the same row state attributes, the corresponding row in the summary table has those attributes. Changing summary table row attributes changes attributes of the corresponding source table rows.

Figure 4.6 Selected Rows in Summary Table Identify Subsets

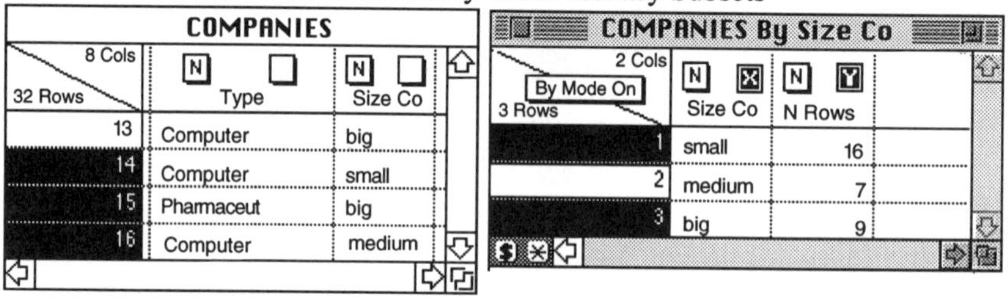

If you use the **Distribution of Y** command in the **Analyze** menu when the summary table is active and its By-Mode is on (**Figure 4.6**), histograms of company type for the selected subsets appear in two separate analysis windows (**Figure 4.7**). In this example the **Set Window Title** command in the **Window** menu is used to modify the window titles.

Figure 4.7 Distributions by Size of Company

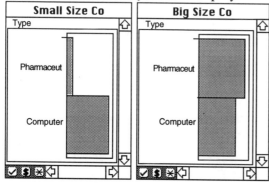

Summary Statistics

Initially, a summary table displays frequency counts (N Rowsa) for each level of the grouping variables. However, you can add columns of descriptive statistics to the table.

The **Stats** pop–up menu in the Group/Summary dialog lists standard univariate descriptive statistics. You can use the **Stats** pop–up menu when you first define a summary table. Or, use the **Add Summary Cols** command in dollar ($) border menu of the summary table at any time (**Figure 4.8**). You can request any of the statistics listed in the **Stats** pop–up menu for any numeric source table column.

Figure 4.8 Add Columns of Summary Statistics

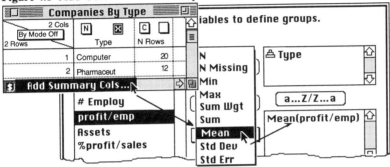

The **Stats** pop–up menu gives these summary statistics for numeric columns:

| N |
| N Missing |
| Min |
| Max |
| Sum Wgt |
| Sum |
| Mean |
| Std Dev |
| Std Err |

N is the number of nonmissing values and is used to compute statistics when there is no column assigned the weight role.

N Missing is the number of missing values.

Min is the least value in a column, excluding missing values.

Max is the greatest value in a column.

Sum Wgt is the sum of all values in a column assigned the Weight (Wgt) role and is used instead of N to compute other statistics.

Sum is the sum of all values in a column.

Mean is the arithmetic average of a column's values. It is the sum of nonmissing values divided by the number of nonmissing values.

Std Dev, the sample standard deviation, is the standard deviation of the non–missing values. It is the square root of the sample variance.

Std Err, the standard error of the mean, is the sample standard deviation, **Std**, divided by the square root of N. If a column is assigned the role of weight, then the denominator is the square root of the sum of the weights.

See **The Moments Report** in Chapter 3, "Simple Regression and Curve Fitting," of the *JMP Statistics and Graphics Guide* for more information about summary statistics.

Follow these steps to add columns of statistics to a summary table:

1. Use the **Add Summary Cols** command in the save ($) menu to again display the Group/Summary dialog (**Figure 4.2**) that shows the **Stats** pop–up menu.
2. Select a numeric column from the source table columns list.
3. Select the statistic you want from the **Stats** pop–up menu.
4. If necessary, repeat steps 2 and 3 to add more statistics to the summary table.
5. Click **Done** to add the columns of statistics to the summary table.

The results illustrated in **Figure 4.9** show the mean the profit/emp in the summary table grouped by type and also in the table grouped by type and size.

Figure 4.9 Expanded Summary tables

N	C Y		N	N	C Y	C Y
Type	N	Mean(profit/emp)	Type	Size Co	N	Mean(profit/emp)
Computer	20	6159.015	Computer	small	14	7998.815
Pharmaceut	12	23546.12	Computer	medium	2	-3462.51
			Computer	big	4	4530.478
			Pharmaceut	small	2	38337.19
			Pharmaceut	medium	5	24035.11
			Pharmaceut	big	5	17140.7

The illustration in **Figure 4.10** shows mean(profit/emp) and mean(Sales($M)) added to the COMPANIES by Type summary table grouped both by Type and Size Co.

Figure 4.10 Summary Statistics for Multiple Source Variables

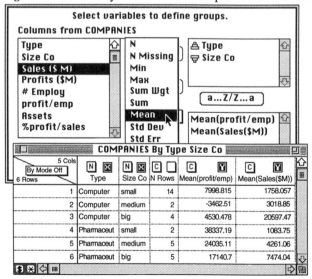

Another way to add summary statistics to a summary table is with the **Subgroup** button on the Group/Summary dialog. This method creates a new column in the summary table for each level of the variable you specify with **Subgroup**. The subgroup variable is usually *nested* within all the grouping variables. The summary table now becomes a two–way table of summary statistics.

The table that groups the COMPANIES data table by Type and Size Co can use a subgroup variable instead of two grouping variables, as illustrated in the next example. **Figure 4.11** begins with the COMPANIES data table grouped by Size Co. The **Stats** pop–up menu requests the mean of Sales($M) with Type as the subgroup variable. This creates a column for each statistic for each level of subgroup variable. The result here is the summary table shown at the bottom of **Figure 4.11**. There is a row for each size company and a column for the mean of each type of company. The cells hold the mean for the subgroup defined by the intersection of the row and column.

Figure 4.11 Summary Statistics for Subgroups

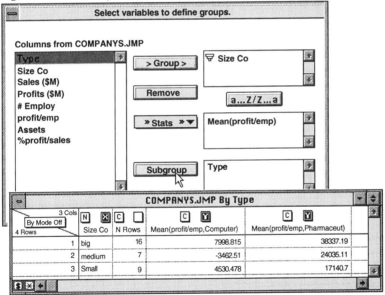

The columns of the Summary table show useful summary information, and are in the form needed by the **Overlay Plot** command in the **Graph** menu.

The Subset Command

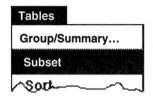

The **Subset** command produces a new data table that consists of selected rows and columns from the active table. To select a column, click the background area above the column name. To select a row, click the row number area in the spreadsheet or define subsets with a summary table as described previously. If no rows or columns are selected, then **Subset** reproduces the entire data table.

Another way to identify subsets is to select bars of a histogram or points on a plot. Click a bar to select it, click or shift–click points in a plot, or use OPTION–click on the Macintosh (ALT–click under Windows) for a stretch rectangle to encompass selected points. All graphical displays that represent the same data table are linked to each other and to the corresponding spreadsheet. Highlighting bars or points in graphical displays is often used to select a range of column values. **Figure 4.12** illustrates selected histogram bars and the corresponding selected rows in the spreadsheet.

After selecting the rows and columns, as shown to the left in **Figure 4.12**, the **Subset** command creates the new data table to the right. This untitled table lists the rows from the original table with the highest values, as highlighted on the histogram.

Figure 4.12 Subsetting Highlighted Columns with a Histogram

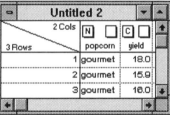

You can also select specific subsets of rows using the **Select Where** command found in the **Rows**. The Where dialog lets you search for a specific value in a column and selects all rows where that value is found. **Where** displays a dialog that prompts you to select a table column to search, a comparison operation from the **Where** dialog pop–up menu, and a selection criterion value. See **The Rows Menu** in Chapter 2, "The Menu Bar," and **Selecting and Deselecting Rows and Columns** in Chapter 3, "JMP Data Tables," for more information about using the **Select Where** command.

The Sort Command

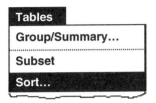

The **Sort** command displays the dialog in **Figure 4.13** to specify columns as sort fields. To sort a JMP data table, select sort fields from the **Columns** list and add them to the **Sort By** list by clicking **Add**. Alternatively, double–click column names to add them to the **Sort By** list, or simply drag them from one list to the other. To remove sort fields, select them in the **Sort By** list and clicking **Remove**.

The columns you add to the **Sort By** list establish the order of precedence for sorting. The first column is the major sort field. Each successive column in the list sorts after the preceding column. The icon to the left of the sort column indicates whether the values sort in ascending or descending order. By default, values sort in ascending order. You can toggle the sort between ascending and descending for any selected column in the **Sort By** list with the **a...Z/Z...a** button. Click **Sort** when the dialog is complete.

Figure 4.13 The Sort Dialog

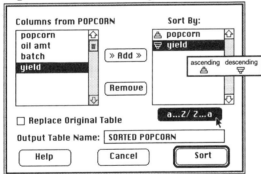

Figure 4.14 shows the POPCORN data table in its original form on the left and sorted by popcorn and yield (within popcorn) on the right. In this example, the **Sort** command created a new data table. However, if you check the **Replace Original Table** option, the sorted table overwrites the original data table.

Figure 4.14 Unsorted and Sorted POPCORN Data Table

POPCORN			SORTED POPCORN		
	popcorn	yield		popcorn	yield
6	gourmet	12.1	6	gourmet	9.2
7	plain	10.6	7	gourmet	8.6
8	gourmet	18.0	8	gourmet	8.2
9	plain	8.8	9	plain	10.6
10	gourmet	8.2	10	plain	10.4
11	plain	8.8	11	plain	10.1

The Stack Columns Command

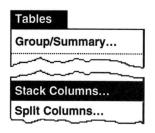

The **Stack Columns** command creates a new data table from the active table by stacking specified columns into a single new column. The new data table preserves the values from the other columns.

Stack Columns creates an *ID* column to identify each row in the new table. The values in that column are the column names from the original table that contained the values now stacked in the new table.

In the example shown in **Figure 4.15**, a JMP data table called POPCORN TRIALS in the SAMPLE DATA folder on the Macintosh (popcrntr.jmp in the data folder under Windows) has two columns that list popcorn yield from two popping trials conducted under various conditions. To complete the appropriate analysis of the data, the two columns called yield1 and yield2, must be *stacked* into a single column.

Figure 4.15 The POPCORN TRIALS Data Table

	popcorn	oil amt	batch	yield1	yield2
1	plain	little	large	8.2	8.8
2	gourmet	little	large	8.6	8.2
3	plain	lots	large	10.4	8.8
4	gourmet	lots	large	9.2	9.8
5	plain	little	small	9.9	10.1
6	gourmet	little	small	2.1	15.9
7	plain	lots	small	10.6	7.4
8	gourmet	lots	small	18.0	16.0

The **Stack Cols** command displays the dialog shown in **Figure 4.16**. To stack columns, select their names from the **Columns** selector list and click **Add**. The column names appear in the **Columns to Stack** list.

This example dialog specifies that the two columns, yield1 and yield2, be stacked into a single column called yield. You can enter any name for the new stacked column in the **Stacked Column Name** box. If you do not enter a name for the new stacked column, its name is _Stacked_ by default. If you delete the default name and do not specify a new name, the new stacked column is not written to the data table.

The **Stack** command creates a special *ID* column to identify each row in the new table. Its values are the column names in the original table from which the stacked values originated. In the example below, the name of the ID column is the default name (_ID_), and it has the literal values 'yield1' or 'yield2.' Be aware that if you delete the default name and do not specify a new name, this new ID column is not written to the data table.

You can enter a name for the data table in the **Output Table** box. This example uses the default table name, **Untitled**.

Figure 4.16 The Stack Dialog

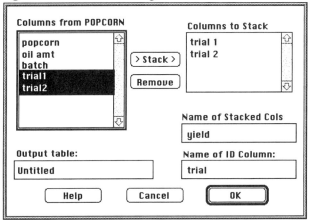

After you complete the dialog, click **Done**. In this example, the stack operation creates the new untitled data table shown in **Figure 4.17**. This new table has twice as many rows as the original table because two columns are stacked. The values of yield are the experimental results, and values in the trial column tell which trial each row represents.

Figure 4.17 Example of a Stacked Column

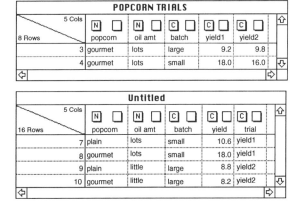

Note➜ You can stack as many columns as needed. However, the number of rows in the new stacked table is the product of the number of stacked columns and the number of rows in the original data table.

The Split Columns Command

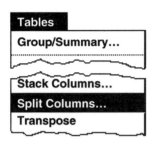

The **Split Columns** command creates a new data table from the active table by dividing a specified column into several new columns according to

- the values (levels) of the Col ID variable
- values (levels) of one or more Grouping variables.

Split Columns displays the dialog shown in **Figure 4.18**. You complete the dialog by specifying columns from the table to be split as follows:

- Select the column or columns whose values are to form multiple new columns and click **Split**.
- Select a single column whose values are to be used as the new column names and click **Col ID**.
- Optionally, select of one or more columns whose values can uniquely identify each row in the new table and click **Group**. If you don't use a grouping variables, **Split** assumes the Col ID values correctly split the rows.
- Use the radio buttons to specify other columns for the new table.

Figure 4.18 The Split Columns Dialog

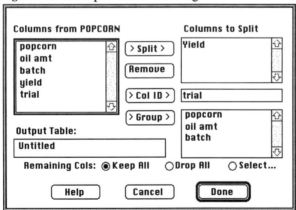

As an example, suppose you want to rearrange the POPCORN (or POPCORN.JMP) table shown at the top of **Figure 4.19** to the form of the **Untitled** table showing beneath it.

The Split Columns dialog (**Figure 4.18**) specifies that the values of the trial column be used as the column names (yield1 and yield2) in the new table. trial is the **Col ID** variable. The values of the new columns, yield1 and yield2, are taken from yield in the original table. yield is the **Split Cols** variable. No **Group** variable is needed in this example. When you click **Done**, you see the untitled table at the bottom of **Figure 4.19**.

Figure 4.19 A Split Columns Example

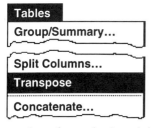

The Transpose Command

The **Transpose** command creates a new JMP table that is the transpose of the active data table. The columns of the active table are the rows of the new table, and its rows are the new table's columns.

The new table has an additional column called Labels whose values are the column names of the original table. If there is no label column, the column names in the transposed table are Row1, Row2, ...Row$n$ where n is the number of rows in the original table. If there is variable assigned the label (Lbl) role, the values of that column are column names in the new transposed table.

The **Transpose** command has the following characteristics:

- The columns of the original table must be either all character or all numeric, except a column used as an ID variable or columns used for grouping.
- **Transpose** can transpose any selected subset of rows.
- **Transpose** can transpose groups of rows. Subsets defined by a summary table created with the **Group/Summary** command, described previously in this section, transpose independently and stack to form the new transposed table.
- Transposing a table with columns but no rows gives a new table with one column that lists those column names. Likewise, if you create a table with one column and assign it the Lbl role, its values become the column names in the transposed table.

A Simple Transpose

The simplest kind of transpose is illustrated in **Figure 4.20**. The example table at the left has two rows and three continuous columns called Plastic, Tin, and Gold. The **Transpose** command creates the untitled table shown to the right. The transposed table has a row for each of the three columns in the example table and columns named Row1 and Row2 for the original table's rows. The additional column called Labels has the column names, Plastic, Tin, and Gold from the Example table as values.

Figure 4.20 Simple Transpose

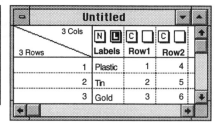

Transpose with a Label

The example table shown in **Figure 4.21** has the same columns as the table used in the previous example and an additional column called Item with the label (Lbl) role. **Transpose** creates the same table as before but uses values in the label column of the Example table as column names in the transposed table.

Figure 4.21 Transpose with Labels

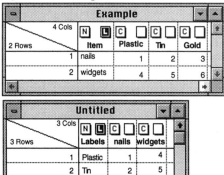

Transpose Groups

Transpose can transpose groups of rows defined with the **Group/Summary** command. Subsets defined by rows in a summary table transpose individually and automatically concatenate to form the new transposed table. The **Group/Summary** command is discussed earlier in this chapter. The tables in **Figure 4.22** illustrate the steps to transpose by groups:

- The left–hand table in **Figure 4.22** has the same columns as those in the two previous examples and an additional variable called year needed for grouping.
- The **Group/Summary** command groups the example table (the *source* table) by year and creates the *summary* table called **Example: by year**, at the right in **Figure 4.22**.
- Highlighting rows in the summary table defines subsets in its source table. **Transpose** treats each subset separately and appends the transposed results to form the final table shown in **Figure 4.23**.

Figure 4.22 An Example Source Table and Its Summary Table

4 Rows \ 5 Cols	year	Item	Plastic	Tin	Gold
1	1990	nails	1	2	3
2	1990	widgets	4	5	6
3	1991	nails	7	8	9
4	1991	widgets	10	11	12

2 Rows \ 2 Cols (By Mode On)	year	N
1	1990	2
2	1991	2

Note➡ You must transpose from an active summary table window to get the results shown. If you make the source table active instead of its summary table, **Transpose** fails to recognize the subsets.

Figure 4.23 Result of Transposing by Groups

Untitled

6 Rows \ 4 Cols	year	Labels	nails	widgets
1	1990	Plastic	1	4
2	1990	Tin	2	5
3	1990	Gold	3	6
4	1991	Plastic	7	10
5	1991	Tin	8	11
6	1991	Gold	9	12

The Concatenate Command

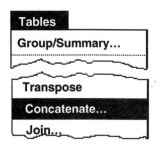

When two or more tables are appended end to end, they are *concatenated*. The **Concatenate** command appends data tables and creates one column in the new table for each column name in the original tables.

If a column name is the same in all tables, the column in the new table lists the values from all tables in the order of concatenation. If a column name is not in all tables, it has missing values for the tables where it does not exist.

Concatenate Tables with the Same Column Names

Suppose you want to concatenate two data tables with the same column names, such as those shown in **Figure 4.24**. Note that the column names do not have to be ordered the same in each table.

Figure 4.24 The TRIAL1 and TRIAL2 Data Tables

TRIAL1				
[N] popcorn	[N] oil amt	[N] batch	[C] yield	[C] trial
1	plain	little	large	8.2	1
2	gourmet	little	large	8.6	1
3	plain	lots	large	10.4	1

TRIAL2				
[C] yield	[C] trial	[N] popcorn	[N] oil amt	[ ] batch
1	8.8	2	plain	little	large
2	8.2	2	gourmet	little	large
3	8.8	2	plain	lots	large

When you select the **Concatenate** command, the dialog shown in **Figure 4.25** lists all the open JMP data tables. To concatenate the data tables above, select TRIAL1 and TRIAL2 from the list on the left and click **Add**.

Figure 4.25 The Concatenate Dialog

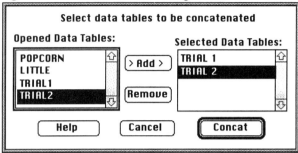

When you click **Concat**, the data tables combine into the new untitled table shown in **Figure 4.26**. All rows from the first data table are followed by all rows from the second data table.

Figure 4.26 Concatenation of the TRIAL1 and TRIAL2 Data Tables

	popcorn	oil amt	batch	yield	trial
1	plain	little	large	8.2	1
2	gourmet	little	large	8.6	1
3	plain	lots	large	10.4	1
4	gourmet	lots	large	9.2	1
5	plain	little	small	9.9	1
6	gourmet	little	small	12.1	1
7	plain	lots	small	10.6	1
8	gourmet	lots	small	18.0	1
9	plain	little	large	8.8	2
10	gourmet	little	large	8.2	2

Concatenate Tables with Different Column Names

Concatenated tables always contains a column for every column name found in the original data tables. For example, even though yield1 in the TRIAL1 table and yield2 in the TRIAL2 table contain similar information (**Figure 4.27**), the new table has columns for both variables. These columns have missing values for rows from the table in which the column did not exist. The column called trial occurs in both tables and concatenates to from a single column in the new table.

Figure 4.27 Data Tables with Different Column Names

TRIAL1				TRIAL2		
N	C	C		N	C	C
batch	yield1	trial		batch	yield2	trial
large	8.2	1		large	8.8	2
large	8.6	1		large	8.2	2
large	10.4	1		large	8.8	2

Figure 4.28 Concatenation When Column Names Are Different

	Untitled					
	N	N	N	C	C	C
	popcorn	oil amt	batch	yield1	trial	yield2
1	plain	little	large	8.2	1	•
2	gourmet	little	large	8.6	1	•
3	plain	lots	large	10.4	1	•
4	gourmet	lots	large	9.2	1	•
5	plain	little	small	9.9	1	•
6	gourmet	little	small	12.1	1	•
7	plain	lots	small	10.6	1	•
8	gourmet	lots	small	18.0	1	•
9	plain	little	large	•	2	8.8
10	gourmet	little	large	•	2	8.2
11	plain	lots	large	•	2	8.8
12	gourmet	lots	large	•	2	9.8
13	plain	little	small	•	2	10.1
14	gourmet	little	small	•	2	15.9
15	plain	lots	small	•	2	7.4
16	gourmet	lots	small	•	2	16.0

You can concatenate as many tables as you want. However, the number of rows in the new table is the sum of the number of rows in all the tables.

The Join Command

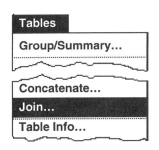

The **Join** command creates a new table by joining two open data tables side by side. Tables can be joined

- by row number
- by matching the values in one or more columns that exist in both data tables
- in a Cartesian fashion where each value in a column of one data table joins with each value in a column of another table to form new rows.

Join by Row Number

The simplest join combines tables by row number. For example, suppose you want to combine the eight rows from each data table shown in **Figure 4.29** into a single table.

This example uses the TRIAL1 and TRIAL2 data from the JMP SAMPLE DATA folder or trial1.jmp and trial2.jmp in the Windows data folder. (The batch and oil amt columns are hidden in some of the Figures that follow.)

Figure 4.29 The TRIAL1 and TRIAL2 Data Tables

TRIAL1			TRIAL2		
	popcorn [N]	yield [C]		popcorn [N]	yield [C]
1	plain	8.2	1	plain	8.8
2	gourmet	8.6	2	gourmet	8.2
3	plain	10.4	3	plain	8.8
4	gourmet	9.2	4	gourmet	9.8
5	plain	9.9	5	plain	10.1
6	gourmet	12.1	6	gourmet	15.9
7	plain	10.6	7	plain	7.4
8	gourmet	18.0	8	gourmet	16.0

When you select **Join**, the dialog shown in **Figure 4.30** displays the name of the active table next to the word **Join**. You select an open data table to join **With** from the table selector list.

Figure 4.30 The Join Dialog

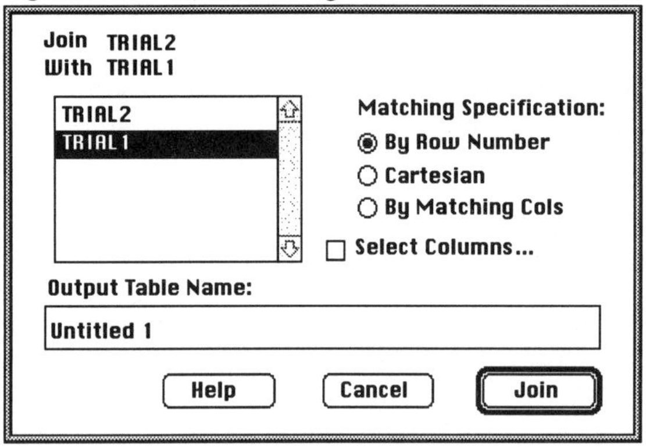

```
Join   TRIAL2
With   TRIAL1

  TRIAL2          Matching Specification:
  TRIAL1          ● By Row Number
                  ○ Cartesian
                  ○ By Matching Cols
                  ☐ Select Columns...

Output Table Name:
Untitled 1

         [ Help ]  [ Cancel ]  [ Join ]
```

After you select a **With** table, choose a matching specification from the dialog as shown in **Figure 4.30**. The default matching option is **By Row Number**. If you match by row number and click **Join**, JMP creates the new data table shown in **Figure 4.31**. The two original tables are joined side by side, and the new table has all columns from both tables.

If a column name is the same in the original tables, the names of these columns in the new table are of the form *columnname:tablename* (yield:TRIAL1 and yield:TRIAL2). If a name (such as oil amt) occurs in only one table, it is written directly to the new table.

Figure 4.31 Result of Join by Row Number

	popcorn:TRIAL1	yield:TRIAL1	popcorn:TRIAL2	yield:TRIAL2
1	plain	8.2	plain	8.8
2	gourmet	8.6	gourmet	8.2
3	plain	10.4	plain	8.8
4	gourmet	9.2	gourmet	9.8
5	plain	9.9	plain	10.1
6	gourmet	12.1	gourmet	15.9
7	plain	10.6	plain	7.4
8	gourmet	18.0	gourmet	16.0

TRIAL1+TRIAL2

Keep a Subset of Columns

If you don't want all columns from the original data tables to be in the joined table, click the **Select Cols** box in the **Join** dialog (**Figure 4.32**) to specify the subset of columns you want. This causes the dialog shown in **Figure 4.33** to show.

Figure 4.32 The Select Columns Option

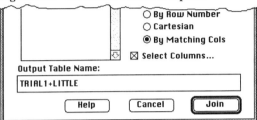

The dialog shows a column selector list for both tables. Select all the columns you want from both tables in the column selector lists and click **Select**. The box under the new table name (TRIAL1+TRIAL2) lists the new column names. In this example, the tables Trial1 and Trial2 (see **Figure 4.29**) have identical data in popcorn column; only one of them is needed in the joined table.

To remove a column from this list, select it and click **Remove**.

Click **Done** when you are ready to return to the **Join** dialog.

Figure 4.33 The Select Columns Dialog

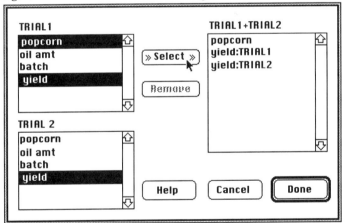

The new table (**Figure 4.34**) has only the selected columns. You can now rearrange columns, rename the columns, change the table name with either the **Table Info** dialog, the **Set Window Name** command, or with the **Save As** command from the **File** menu.

Figure 4.34 Joined Table with Column Selection

	popcorn	yield:TRIAL1	yield:TRIAL2
1	plain	8.2	8.8
2	gourmet	8.6	8.2
3	plain	10.4	8.8
4	gourmet	9.2	9.8
5	plain	9.9	10.1
6	gourmet	12.1	15.9
7	plain	10.6	7.4
8	gourmet	18.0	16.0

Join by Matching Columns

Often you need to join two tables that have different numbers of rows. This means that columns with similar information do not have all matching values. For example, suppose the popcorn trials experiment is complete for the first trial but partially complete for the second trial. The TRIAL1 data table, shown in **Figure 4.35**, has values for eight experimental conditions. The LITTLE table has values for only four of those conditions. Further, columns having similar information in the two tables do not have the same names (oil amt and oil) and are not in the same relative positions in the data tables.

Figure 4.35 Joined Tables with Different Columns

TRIAL1

	popcorn	oil amt	batch	yield
1	plain	little	large	8.2
2	gourmet	little	large	8.6
3	plain	lots	large	10.4
4	gourmet	lots	large	9.2
5	plain	little	small	9.9
6	gou			
7	plai			
8	gou			

LITTLE

	popcorn	yield	batch	oil
1	plain	8.8	large	little
2	gourmet	8.2	large	little
3	plain	10.1	small	little
4	gourmet	15.9	small	little

To join these tables correctly, the values for **popcorn** and **batch** must match in both tables, and **oil amt** in the TRIAL1 table must match with **oil** in the LITTLE table. The **By Matching Cols** option at the bottom of the Join dialog (**Figure 4.36**) lets you specify columns whose values must match to complete a join.

Figure 4.36 Join By Matching Cols Option

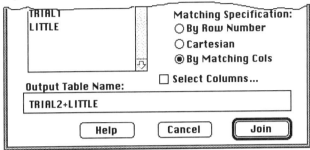

By **Matching Cols** gives the dialog shown in **Figure 4.37** that prompts you to select columns whose values must match in both tables for rows to be joined. After selecting a column from each upper list, click **Match**. This displays the selected pair of columns in the lower boxes. Select additional pairs of columns as needed. If you want to remove matched columns, select them in the lower lists and click **Remove**.

Matching columns do not have to have the same names and do not have to be in the same relative column position in both tables. When you click **Done**, the first column in the left–hand list pairs with the first column in the right–hand list. Likewise the second columns are paired, and so on. Rows join only if values match for all the column pairs.

Figure 4.37 The Specify Matching Columns Dialog

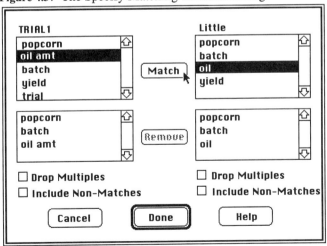

After you choose columns whose values must match, additional options show beneath each table's column list.

Figure 4.38 Choose Matching Options

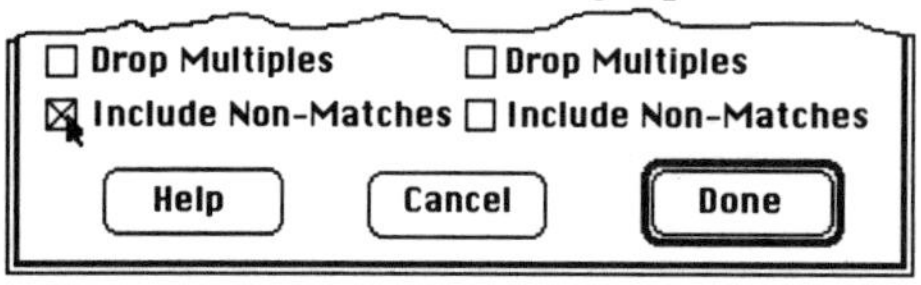

For example, some rows in the TRIAL1 table (see **Figure 4.35**) do not have a match in the LITTLE table. If you don't check the **Include Non–Matches** box for the TRIAL1 table, those columns do not appear in the new table. Click **Include Non–Matches** box for the TRIAL1 table so that the new table, called TRIAL1+LITTLE, has all rows in TRIAL1 even if there is no match in the LITTLE table.

After the Specify Matching Columns dialog is complete, click **Done** to return to the Join dialog. When you click **Join**, JMP creates the table shown in **Figure 4.39**. The yield column from the LITTLE data table (Yield:LITTLE) has missing values whenever there were no matching values in the TRIAL1 table.

Notice that the new table is now sorted by the matching columns. JMP automatically sorts the data table so that matching takes place properly. You do not need to sort tables before joining them with matched columns.

Figure 4.39 Joined Table with Matched Columns and Selected Variables

8 Rows / 5 Cols	popcorn:...	oil mat:...	batch:...	yield:TRIAL 1	yield:LITTLE
1	gourmet	little	large	8.6	8.2
2	gourmet	little	small	12.1	15.9
3	gourmet	lots	large	9.2	•
4	gourmet	lots	small	18.0	•
5	plain	little	large	8.2	8.8
6	plain	little	small	9.9	10.1
7	plain	lots	large	10.4	•
8	plain	lots	small	10.6	•

There are additional **Join** options:

Drop Multiples

You can specify **Drop Multiples** for either or both of the data tables. If you drop multiples in both tables, only the first match found is written to the new table. If you specify this option for one table, the first match value is joined with all matches in the other table.

Include Non-Matches

When you select **Include Non-Matches** for a data table, each row from that data table is included in the new data table even when is no matching value, as in the previous example. You can specify this option for either or both data tables being joined.

If you do not select either the **Drop Multiples** or the **Include Non-Matches** option, a Cartesian join is performed within each group of matching column values.

Cartesian Join

If you choose the **Cartesian** join option, each row in the **Join** data table joins with each row in the **With** data table. To illustrate this, suppose you want to construct a JMP table that has a row for each combination of levels of experimental conditions in the popcorn example. You can begin with the three small tables shown in **Figure 4.40**. Each table has two rows and one column. The values are the experimental categories for popcorn yield trials.

Figure 4.40 JMP Tables for a Cartesian Join

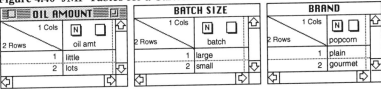

The **Join** command joins the active data table with any other open data table you choose. In this example, you must use the **Join** command twice. The first join (see **Figures 4.41** and **4.42**) combines the OIL AMOUNT data table with the BATCH SIZE table using the Cartesian option.

Figure 4.41 A Cartesian Join

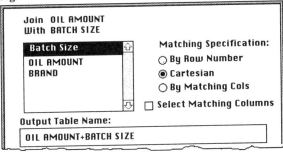

The joined data table shown in **Figure 4.42** has all columns from the original tables. Each value in the OIL AMOUNT table pairs with each value in the BATCH SIZE table, giving a new table with four rows.

Figure 4.42 Cartesian Join Example

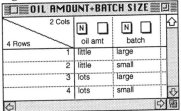

Next, make the BRAND table active and **Join** again as shown in **Figure 4.43**. Select the new table, OIL AMOUNT+BATCH SIZE, as the **With** table, type POPCORN as the **Output Table Name**, use the **Cartesian** option, and click **Join** to produce the table in **Figure 4.44**.

Figure 4.43 Cartesian Join Dialog

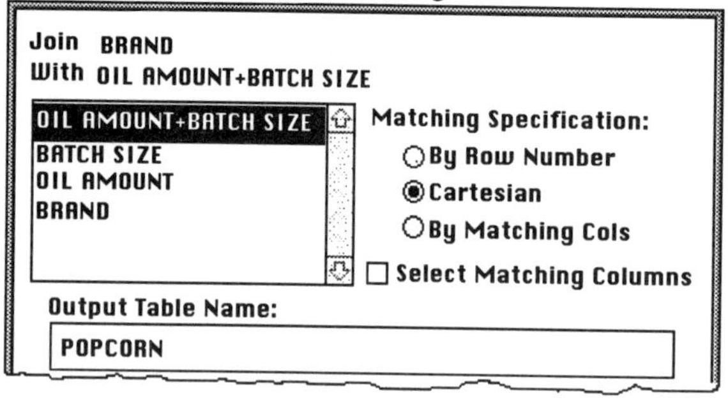

This final data table has a row for each experiment condition. The new POPCORN data table is ready for recording results of the corn popping trials.

Figure 4.44 Final Table from Cartesian Join Example

POPCORN

3 Cols / 8 Rows	popcorn	oil amt	batch
1	plain	little	large
2	plain	little	small
3	plain	lots	large
4	plain	lots	small
5	gourmet	little	large
6	gourmet	little	small
7	gourmet	lots	large
8	gourmet	lots	small

Keep in mind that the number of rows produced by a Cartesian join is the product of the number of rows in the two original tables.

The Table Info Command

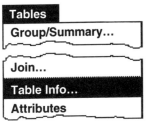

The **Table Info** command displays the dialog shown in Figure 4.45. You can use this dialog to make notes about the current data table or to examine its attributes.

You can also lock the data table if you have not modified it since last saving it. Making changes in the data table dims the **Lock** box, and the **Dirty** box is checked.

To change the data table name, type a new name into the dialog. This renames the copy of the table in memory, but the copy on disk remains as last saved. The new table is not saved until you select the **Save** or **Save As** command.

Figure 4.45 The Table Info Dialog

The Attributes Command

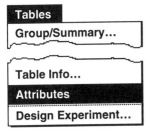

The **attributes** command creates a new table called an *attributes table* from the active data table, called its *source table*.

- An attributes table has a row for each column in its source table.
- All attributes tables have a column for each type of column characteristic.

The attributes table is linked to its source table. You can modify the characteristics of the source table columns by editing values in corresponding attributes table rows. When you choose **Update Source** in the dollar ($) pop-up menu of the attributes table, the source table updates to include the changes. An advantage of using an attributes table is that you can change the characteristics of many source table columns at the same time instead of using Column Info dialogs and changing characteristics one column at a time.

The Table Info button in the upper left corner of the attributes table accesses the Table Info dialog of its source table.

Figure 4.46 A JMP Table and Its Attributes Table

	POPCORN.JMP				
5 Cols	N	N	N	C	C
16 Rows	popcorn	oil amt	batch	yield	trial
7	plain	lots	small	10.6	1
8	gourmet	lots	small	18.0	1
9	plain	little	large	8.8	2
10	gourmet	little	large	8.2	2

	Attributes of POPCORN.JMP								
9 Cols / Table Info	N	L	N	N	C	C	N	N	N
5 Rows	Name	Type	Measure	FW	N Dec	Lock	Source	Validation	Role
1	popcorn	Character	Nominal	7	?	No	No Formula	None	None
2	oil amt	Character	Nominal	6	?	No	No Formula	None	None
3	batch	Character	Nominal	6	?	No	No Formula	None	None
4	yield	Numeric	Interval	6	1	No	No Formula	None	None
5	trial	Numeric	Interval	5	0	No	No Formula	None	None

Attributes Table Columns

The attributes tables has a column for each type of column characteristic. Column characteristics are described briefly here. You can find other discussions in **The Tables Menu** section of Chapter 2, "The Menu Bar," and in Chapter 3, "JMP Data Tables."

Name

The Name column lists the variable names in the source table. You type in new column names using up to 31 keyboard characters. If you enter a long name, expand the column width of both tables by dragging the column boundaries.

Type

The Type column has values Numeric, Character, or Row State. You can change a column's type keeping in mind these considerations:

- Numeric columns must contain numbers as values in a source table. If you change a column type from character to numeric in an attributes table and update its source table, any character values in the source table become missing values. If you change the type to Row State, numbers are treated as codes and converted to the corresponding Row States and characters become null row states.

- Character columns can contain any characters including numbers. If you change a column type from numeric to character in an attributes table and update its source

table, numbers are treated as discrete values instead of continuous values. If you change the type to Row State, numbers are treated as codes and converted to the corresponding Row States. Characters become null row states.
- Row State columns contain special information that can affect the appearance of plots. If you change a column type from Row State to Numeric or Character and update the source table, the results are number codes.

Measure

The Measure column has values Continuous, Ordinal, and Nominal, which specify the modeling type JMP uses for data analysis and plotting. You can change a column's modeling type as long as it remains compatible with its data type.

Numeric columns can be continuous, ordinal, or nominal.

Character columns can be only nominal or ordinal.

Row State columns have no modeling type.

Data types and modeling types are described further in the **Cols Menu** section of this chapter and in Chapter 3, "JMP Data Tables."

FW

The FW column lists the field width of each column. The maximum field width is 40 for numeric values and is 255 for character values. Truncation occurs in the source table if you change a column's field width so that it is too small to contain its values.

N Dec

The N Dec column lists the number of decimal places assigned to numeric columns. The default number of decimal places is 8. Changing the number of places in the attributes table is the same as using the **Format** pop–up menu in the Column Info dialog.

Lock

The Lock column has values Yes or No, which tell whether a source table column is locked or unlocked. A locked column renders it *uneditable*. When you use a formula to compute values for the column, it is automatically locked. You can also click the Column Info **Lock** box to protect any column's values.

Source

The Source column has values No Formula, Formula, or Instrument to indicate the source of the data. A typical use of Source and Lock is to remove a column's formula and locked status when a formula generated data, but you want to use the result as a free column.

Validation

The Validation column has values None, Range Check, or List Check to indicate admissible values in a column. You can use the **Update Col Check List** on the Column Info dialog to add or delete validation values.

Role

The Role column has one of the role assignment pop–menu selections as its value. You can change role assignments by changing these values, but only one column can have the Label role. If you change a role value to Label when another label column exists, the previous label column's value automatically becomes None.

Adding, Deleting, and Moving Source Table Columns

You can add, delete, and reposition columns in a source table by rearranging the corresponding rows in its attributes table. An attributes table responds to commands in the **Rows** and **Cols** menus. If you move rows in an attributes table, the columns in its source table reposition accordingly when you update it. You can add columns to a source table by adding rows to its attributes table. Likewise, deleting rows in an Attribute deletes the corresponding columns in its source table when you update it.

You cannot delete columns the attributes table that can affect the source table. However, you can move columns in an attributes table for your convenience. Adding columns to an attributes table has no effect on a source table when you update it.

You can also rearrange source table columns with the **Sort** command in the attributes table dollar ($) menu, described in the next section.

Updating a Source Table

At any time while you are editing an attributes table, you can update its source table by selecting the **Update Source** command in the dollar ($) pop-up menu at the lower left of the attributes table. Other commands in the dollar menu are as follows:

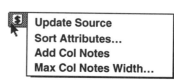

Sort Attributes

accesses the Sort dialog used by the **Sort** command. You can sort the attributes table by any of its variables. When you update the source table. the columns rearrange according to the sort you requested.

Add Col Notes

adds a character column to the attributes table called Col Notes. By default, it has a length of 64. You can change the length of this column using its Column Info dialog. If there are any currently existing column notes, they appear in the new Col Notes column. Any entry you make into the Col Notes column appears in the Column Info dialog for the corresponding source table column after you update it.

Max Col Note Width

lets you specify a maximum width to allow for each column notes up to a maximum width of 255.

The Design Experiment Command

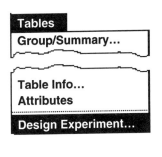

The **Design Experiment** command can creates five kinds of experimental designs:

- two–level designs for screening factors
- response surface designs to focus on optimal values for a set of continuous factors.
- mixed–level designs to detect quadratic effects of some factors.
- mixture designs to find best relative proportions of ingredients for a mixture.
- general factorial design for all combination of factor levels

Design Experiment creates the table of runs for the design you specify. After completing the experiment it is then convenient to use JMP to analyze the results.

See Chapters 25 through 30 in the *JMP Statistics and Graphics Guide* for complete documentation of this command.

Chapter 5
Calculator Functions

The JMP calculator is a powerful tool for building formulas that calculate column values. JMP formulas can use information from existing columns in the data table, built–in JMP functions, and constants. Formulas can be simple assignments of numeric, character, or row state constants or can contain complex evaluations based on conditional clauses.

When you create a formula for a column, that formula becomes an integral part of the data table. The formula is stored as part of a column's information when you save the data table, and it is retrieved when you reopen the data table. You can examine or change a column's formula at any time by opening the calculator window.

A column whose values are computed by a formula is both *linked* and *locked*. It is linked to (or *dependent on*) all other columns that are part of its formula. Its values are automatically recomputed whenever you edit the values in these columns. It is also locked so that its data values cannot be edited, which would invalidate its formula.

The calculator window operates like a pocket calculator with buttons, displays, and an extensive list of easy–to–use features. This chapter begins with a simple calculator example and serves as a reference guide for all calculator functions and features. Chapter 6, "Using the Calculator," shows you how to create and use formulas.

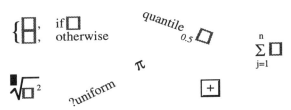

Chapter 5
Contents

A Quick Example 133
The Calculator Window 135
Calculator Terminology 136
The Calculator Work Panel 138
The Formula Display 139
Keypad Functions 140
Function Browser Definitions 141
    Terms Functions 143
    Numeric Functions 144
    Transcendental Functions 146
    Character Functions 149
    Comparisons 153
    Conditions 154
    Random Number Functions 159
    Probability Functions 160
    Statistical Functions 163
    Date Functions 168
    Row State Functions 170
    Parameters 173
    Variables 174
    Editing Functions 175

A Quick Example

The following example gives you a quick look at the basic features of the calculator. If you prefer a hands–on approach to learning software, you can review this example and then explore the calculator on your own. Keypad and function browser definitions are in this chapter. Chapter 6, "Using the Calculator," gives details about building and changing formulas, and shows practical examples.

For this example, open the STUDENTS data table in the SAMPLE DATA folder (students.jmp in the Windows data folder). It has a column called weight. Suppose you want a new column that has computed standardized weight values. Begin by selecting **New Column** from the **Cols** menu.

Figure 5.1 The Cols Menu

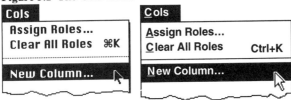

The dialog shown in **Figure 5.2** lets you set the new column's characteristics. Type the new name, Std. Weight, in the **Col Name** area and click **Data Source: Formula**. The other column characteristics define a numeric continuous variable and are correct for this example.

Figure 5.2 The New Column Dialog

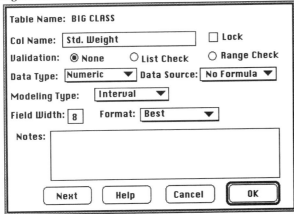

When you select **Formula** from the **Data Source** pop–up menu and click **OK**, the calculator window opens as shown in **Figure 5.3**.

Figure 5.3 The Calculator Window

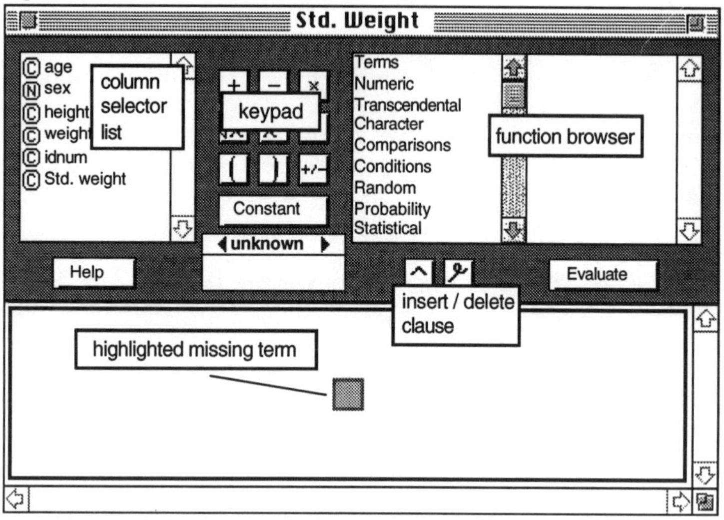

Next, enter the formula that standardizes the weight values by following these steps.

1. While the initial missing term is highlighted, click on the column named **weight** in the column selector list.

   weight

2. Click on the minus button in the keypad.

   weight − ■

3. While the new missing term is highlighted, click on **weight** again.

   weight − weight

4. Click in the function browser topics list and scroll down the topics to locate **Statistical** functions. Click on this topic to see a list of statistical functions on the right half of the function browser. Then click **Mean** in this list.

   weight − weight

5. Click on the right parenthesis in the keypad to highlight the entire expression.

   weight − weight

6. While the entire expression is highlighted, click on the division button in the keypad.

$$\frac{(weight-\overline{weight})}{\blacksquare}$$

7. Choose **weight** again from the column selector list.

$$\frac{(weight-\overline{weight})}{weight}$$

8. While **weight** is still highlighted in the denominator, choose **Std. Deviation** from the right half of the function browser.

$$\frac{(weight-\overline{weight})}{\text{std } weight}$$

You have now entered your first formula. Close the calculator window to see the new column fill with values. If you change any of the **weight** values, the calculated **std. weight** values change automatically.

What if you made a mistake?

If you make a mistake entering a formula, first choose **Undo** from the **Edit** menu. **Undo** reverses the effect of the last command. There are other editing commands to help you modify formulas, including **Cut**, **Copy** and **Paste**, and the delete key to remove selected expressions. If you need to rearrange terms or expressions, you can use the hand tool to move or swap formula pieces. Editing formulas and using the hand tool are discussed later in this chapter.

The Calculator Window

The JMP calculator is a window that computes values for a column. You can open a column's calculator window three ways:

- Select **Formula** from the **Data Source** pop-up menu in a New Column dialog, and then click **OK**.
- Click the formula display field in the Column Info dialog of an existing column.
- Option–Click in the white space at the top of a column that has a formula to open its calculator without first opening the Column Info dialog.

The two main areas of the calculator window are called the *calculator work panel* and the *formula display,* as illustrated by **Figure 5.4**.

- The calculator work panel is composed of buttons (help, constant, add and delete, and Evaluate), selector lists, and a constant entry field.
- The formula display is an editing area you use to construct and modify formulas.

Figure 5.4 The Calculator Window

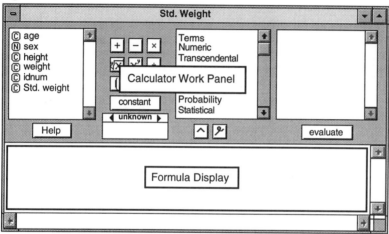

The sections that follow define calculator terminology and give the details you need to use the calculator work panel and the formula display.

Calculator Terminology

The following list is a glossary of terminology used throughout this chapter and in Chapter 6, "Using the Calculator". Each italicized word within a definition is also defined.

argument

A constant, column, or temporary variable *expression* (including mathematical operands) that is operated on by a *function*.

term
   An indivisible part of an expression. Constants and variables are terms.

    ← term

expression
   A formula or any part of it that can be highlighted as a single unit, including *terms, missing terms,* and *functions* grouped with their *arguments*.

   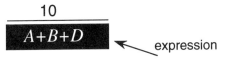 ← expression

function
   A mathematical or logical operation that performs a specific action on one or more *arguments*. Functions include most items in the function browser and all keypad operators.

   function →

missing term
   Any empty place holder for an *expression,* represented by the small box shown below.

    ← missing term

missing value
   Excluded or null data consisting of the missing value mark (a period on the Macintosh and ? under Windows) for numeric data, or null character strings for character data.

   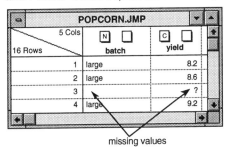
   missing values

The Calculator Work Panel

The calculator work panel is composed of buttons and selection lists as illustrated in **Figure 5.5**.

Figure 5.5 The Calculator Panel

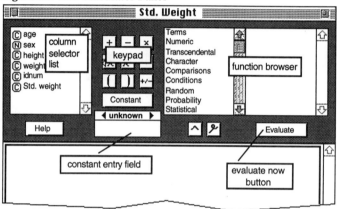

The following paragraphs describe each component of the calculator work panel:

The Column Selector List
   displays all columns from the current data table. To choose a column, highlight an expression in the formula and click the appropriate column name.

The Keypad
   is a set of buttons for commonly used calculator functions.

The Function Browser
   groups collections of calculator functions and features in lists organized by topic. To enter a function in a formula, highlight an expression and click any item in one of the function browser topics.

The Help Button
   displays a help window that provides information to help you use the calculator.

The Constant Entry Field
   is an editable field used to enter any type of constant. The bar above this field identifies the current data type, and the triangles on either side of the bar scroll to other data types. You can automatically switch the data type to character by preceding your entry with a quotation mark.

   To enter a constant, highlight an expression in the formula and scroll to the appropriate data type. The correct data type often displays automatically based on the data type of the highlighted expression. After entering the constant, click **Constant** or type the RETURN key to enter the value in the formula.

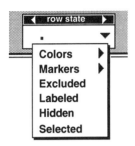

If you specify a constant as row state data, the default row state constant (the black dot) appears in the constant entry field. Click this field to access the pop-up menu shown to the left, with options for all the row states. You can use this pop-up menu to modify the row state constant.

The Constant Button
enters the current contents of the constant entry field in the formula, replacing the highlighted expression.

The Insert and Delete Clause Buttons
are located to the right of the constant entry field. A clause is a set of arguments to a function that accepts a varying number of arguments.

 inserts a new clause into a formula.

 deletes an existing clause from a formula.

To insert a clause into a formula, first select the existing clause that you want the new clause to follow. When you click the insert button, the new highlighted clause appears. To delete a clause from a formula, select it and click the delete button. This deletes the clause and highlights the following clause.

See the sections **Numeric Functions**, **Conditions**, and **Row State** in this chapter for details about using clauses.

The Evaluate Button
automatically fills spreadsheet columns with calculated values whenever you change a formula and then close the calculator window or make it inactive. Use the Evaluate button to calculate a column's values while the calculator window remains active.

The Formula Display

The formula display is the area where you build and view a formula. To compose a formula, highlight expressions in the formula display and apply functions and terms from the calculator panel.

Functions always *operate* upon highlighted expressions, terms always *replace* highlighted expressions, and arguments are always *grouped* with functions. To find which expressions serve as a function's arguments, highlight that function. These groupings also show how order of precedence rules apply and show which arguments will be deleted if you delete a function. Chapter 6, "Using the Calculator," tells you how to build and use formulas.

Keypad Functions

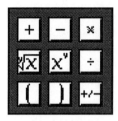

The keypad is composed of common operators (referred to as functions in all calculator documentation). Enter a keypad function by highlighting an expression in the formula display and clicking on the appropriate keypad button or by entering a keyboard shortcut. See the section **Keyboard Shortcuts** in Chapter 6, "Using the Calculator," for a list of keyboard equivalents.

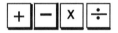 **Arithmetic Functions**

These four arithmetic functions work as they normally do on a pocket calculator and build expressions like this addition function:

Column1+Column2

 Radical Function

The radical function calculates the specified root of a radicand. It has an implied index of 2 (a square root), which is not displayed. The index area is highlighted as shown below so that you can enter a different value. Only indexes other than 2 are displayed.

 General Exponential Function

The general exponential function raises a given value to a specified power. It has an exponent of two by default. The power is highlighted as shown below and can be changed to another value.

 Apply Parentheses Function

The left parenthesis key applies parentheses around a highlighted expression. The applied parentheses are a special function that controls the order in which an expression is evaluated. The example below includes parentheses that force the addition of *Column 1* and *Column 2* before multiplication:

Column1 • (*Column1* + *Column2*)

Highlight Expression Function

The right parenthesis key highlights an expression, starting with the current expression and expanding outward each time the key is clicked. If no parentheses exist further outward, the right parenthesis highlights the entire formula. The right parenthesis also stops at items that act like parentheses such as the absolute value sign and the floor and ceiling signs.

In the following example, Column1 is highlighted and then the right parenthesis button is clicked twice:

$$\left(\frac{Column1 \bullet (\boxed{Column1} + Column2)}{10}\right)$$

$$\left(\frac{Column1 \bullet \boxed{(Column1 + Column2)}}{10}\right)$$

$$\boxed{\left(\frac{Column1 \bullet (Column1 + Column2)}{10}\right)}$$

 **Unary Sign Function**

The unary sign function inverts the sign of its argument. Apply the function to variable expressions or use it to enter negative constants as shown below:

5 becomes -5

Function Browser Definitions

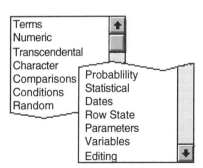

The *function browser* groups the calculator functions by the topics shown at left. To enter a function, highlight an expression and click any item in the function browser topics or use the function's keyboard shortcut. See the section **Keyboard Shortcuts** in Chapter 6, "Using the Calculator," for a list of keyboard equivalents.

The function categories are briefly described in the following list. The remaining sections in this chapter give a detailed description and examples of each function in the function categories. They are presented in the order you find them in the function browser.

- **Terms** list commonly needed constants or functions such as e, π, current row number, assigning values, or using a subscript.

- **Numeric** lists functions such as Round, Absolute Value, Maximum, and Minimum.

- **Transcendental** are the standard transcendental functions such as natural and common log, sine, cosine, tangent, inverse functions, and hyperbolic functions.

- **Character** lists functions that operate on character arguments for trimming, finding the length of a string, and changing numbers to characters or characters to numbers.

- **Comparisons** are the standard logical comparisons such as less than, less than or equal to, not equal to, and so forth.

- **Conditions** are the logical functions Not, And, and Or. They also include programming-like functions such as Assign, If/Otherwise, Match, and Choose.

- **Random** functions generate uniform or normal random numbers. There is also a function to randomize the order of table rows.

- **Probability** functions compute probabilities and quantiles for the normal, Student's t, chi-square, and F distributions.

- **Statistical** lists a variety of functions that calculate standard statistical quantities such as the mean or standard deviation.

- **Dates** require arguments with the *date* data type, which is interpreted as the number of seconds since January 1, 1904. Date functions return values such as day, week, or month of the year, compute dates, and can find data intervals.

- **Row State** functions assign or detect row state status of color, marker, label, hidden, excluded, or selected.

- **Parameters** are named constants that you create and can use in any formula.

- **Variables** are named, temporary variables that you assign an expression and use in other expressions for a given column.

- **Editing** functions let you alter the size of the font used in the formula display, replace a formula with its derivative, or discard any changes made to a formula.

Terms Functions

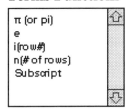

The Terms selector list consists of terms commonly used in formulas. To select from this list, highlight an expression and click a term. The following list describes the Terms:

π (written as pi under Windows)
    is the numeric constant pi, 3.14159265....

e
    is the numeric base of natural logarithms, 2.7182818....

i (row #)
    is the current row number when an expression is evaluated for that row. You can incorporate i in any expression, including those used as column name subscripts. The default subscript of a column name is i unless otherwise specified.

n (# of rows)
    is the total number of rows in the active data table.

Subscript
    enables you to use a column's value from a row other than the current row. Highlight a column name in the formula display and click **subscript** to display the column's default subscript i. The i highlights and can be changed to any numeric expression. Subscripts that evaluate to nonexistent row numbers produce missing values. A column name without a subscript refers to the current row.

To remove a subscript from a column, select the subscript and delete it. Then delete the missing box.

The formula

$$Count_i - Count_{i-1}$$

uses subscripts to calculate the difference between each pair of values from the column named **count**. When i is 1, the computation produces a missing value.

The formula

$$\begin{cases} 1, & \text{if } i \leq 2 \\ fib_{i-1} + fib_{i-2} & \text{otherwise} \end{cases}$$

calculates a column called fib, which contains the terms of the *Fibonacci* series (each value is the sum of the two preceding values in the calculated column). It shows the use of subscripts to do recursive calculations. A recursive formula includes the name of the calculated column, subscripted such that it references only previously evaluated rows (rows 1 through i–1).

The calculation of the Fibonacci series shown above includes a conditional expression and a comparison. See the sections called **Conditions** and **Comparisons** for more information about these function browser topics.

Numeric Functions

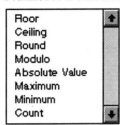

Numeric functions require numeric values as arguments and calculate numeric values as results. If you have numbers stored as characters and want to use them in a numeric function, first convert them to numbers with the num-to-char function found in the list of character functions (num ☐).

Each numeric function is described as follows:

The Floor function returns the largest integer less than or equal to its argument. For example,

$\lfloor 2.7 \rfloor$ results in 2, while $\lfloor -0.5 \rfloor$ results in –1.

The Ceiling function returns the smallest integer greater than or equal to its argument. For example,

$\lceil 2.3 \rceil$ results in 3, while $\lceil -2.3 \rceil$ results in –2.

round(☐, to 0 decimals)

The Round function rounds the first argument to the number of decimal places given by the second argument. For example,

round(3.554, to 2 decimals) rounds to 3.55 and
round(3.555, to 2 decimals) rounds to 3.56.

☐ mod ☐

The Modulo function returns the remainder when the second argument is divided into the first. For example,

6 mod 5 results in 1.

|☐|

The Absolute Value function returns a positive number of the same magnitude as the value of its argument. For example, both

|5| and |-5| both result in 5.

max(☐)

The Maximum function takes the maximum of its numeric or character arguments. Max ignores missing values. For example, the expression

max(3, 7, •) results in 3. Use [^] and [✱] in the calculator panel to add blank Maximum function arguments or remove unwanted arguments.

min(☐)
: The Minimum function takes the minimum of its numeric or character arguments. Min ignores missing values. For example, the expression

> min(3, 7, •) results in –7. Use ⌃ and 🔧 in the calculator panel to add blank

Minimum function arguments or remove unwanted arguments.

count(from ☐, to ☐, in ☐ steps, 1 time)
: The count function creates a list of values beginning with the from value and ending with the to value. The number of steps specifies the number of values in the list between and including the from and to values. Each value determined by the first three arguments of the count function occurs consecutively the number of times you specify with the time argument. When the to value is reached, count starts over at the from value.

For example, the columns in the data table shown in **Figure 5.6** result from the following formulas:

> count1 is count(from 1, to 9, in 2 steps, 1 time)
> count2 is count(from 1, to 9, in 3 steps, 1 time)
> count3 is count(from 1, to 9, in 9 steps, 1 time)
> count4 is count(from 1, to 9, in 3 steps, 3 times)

Figure 5.6 Examples of the Count Function

	Count1	Count 2	Count 3	Count 4
1	1	1	9	1
2	9	5	9	1
3	1	9	9	1
4	9	1	9	5
5	1	5	9	5
6	9	9	9	5
7	1	1	9	9
8	9	5	9	9
9	1	9	9	9

The count function is especially useful for generating a columns of grid values. For example, the following formulas create a square grid of increment n (the term function for number of rows) and axes that range from –5 to 5. There are table templates in the SAMPLE DATA folder for grids. See Chapter 6, "Using the Calculator." for details.

> count(from -5, to 5, in $\sqrt{n}$ steps, 1 time)
> count(from -5, to 5, in $\sqrt{n}$ steps, $\sqrt{n}$ times)

Transcendental Functions

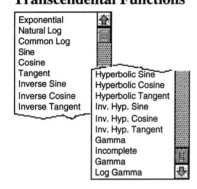

The calculator supports logarithmic functions for any base and selected trigonometric functions. The following list shows how each function appears in the calculator, defines the function, and gives an example:

$e^{\square}$

The Exponential function raises the constant e to the specified power. It is the same as applying the general exponential key to the term e. The expression

e^{-3} evaluates as 0.0498.

ln $\square$

The Natural Log function calculates the natural logarithm (base e) of its argument. The expression,

ln e evaluates as 1.

log $_{10}\square$

The Common Log function calculates a common logarithm (base 10). You can change the default log base from 10 to another value using the Constant field. Both log arguments can be numeric expressions. The expression

log $_2$ 32 evaluates as 5.

sin $\square$ and cos $\square$

The Sine and Cosine functions calculate the sine and cosine of their respective arguments given in radians. For example, the expressions

sin 0 evaluates as 0 and cos 0 evaluates as 1.

tan $\square$

The Tangent function calculates the tangent of an argument given in radians. The expression

$\tan\left(\dfrac{\pi}{4}\right)$ evaluates as 1.

arcsin $\square$ and arccos $\square$

The Inverse Sine and Inverse Cosine functions return the inverse sine (arcsine) and inverse cosine (arccosine) of their respective arguments. The returned value is measured in radians. For example, both expressions

arcsin 1 and arccos 0 evaluate as 1.57080.

arctan ☐

The Inverse Tangent function returns the inverse tangent (arctangent) of its argument. The returned value is measured in radians. The expression

arctan 0.5 evaluates as 0.46364.

sinh ☐ and cosh ☐

The Hyperbolic Sine and Hyperbolic Cosine functions return the hyperbolic sine and hyperbolic cosine of their respective arguments. The expression

sinh 1 evaluates as 1.175201, and cosh 0 evaluates as 1.0.

tanh ☐

The Hyperbolic Tangent function returns the hyperbolic tangent of its argument. The expression

tanh 1 evaluates as 0.761594.

arcsinh ☐ and arccosh ☐

The Inverse Hyperbolic Sine and Inverse Hyperbolic Cosine functions return the inverse hyperbolic sine and inverse hyperbolic cosine of their respective arguments. The expression

arcsinh 1 evaluates as 0.881374, and arccosh 1 is 0.

arctanh ☐

The Inverse Hyperbolic Tangent function returns the inverse hyperbolic tangent of its argument. The expression

arctanh 0.5 evaluates as 0.549306.

Γ(☐)

The Gamma function, denoted by Γ(i) is defined by

$$\Gamma(i) = \int_0^\infty x^{i-1} e^{-x} dx$$

In JMP, this formula computes gamma for each row using i, which is the current row number. It is used in the formula for the gamma distribution and other probability distributions. Other interesting gamma function relationships are

$$\Gamma(i+1) = i \bullet \Gamma(i) = i \text{ factorial (denoted } i!\text{), and } \Gamma(0.5) = \sqrt{\pi}$$

The two parameter Beta function, B(m, n), is written terms of the Gamma function as

$$B(m,n) = \frac{\Gamma(m)\Gamma(n)}{\Gamma(m+n)}$$

igamma(☐, shape 1)

The **Incomplete Gamma** function, is the same as **Gamma** except the integral is incomplete and has a second parameter called shape in the JMP formula. The incomplete gamma is written

$$\text{igamma}(a, i) = \frac{1}{\Gamma(a)} \int_0^i x^{a-1} e^{-x} dx$$

The JMP formula

igamma(i, shape 1)

uses the default shape parameter (1) and computes the incomplete gamma for each row using the row number *i.*, which is the limit of the integral.

lgamma(☐)

The **Log Gamma** function is the natural log of the result of the gamma function evaluation. You get the same result using the natural log function in the calculator with the Gamma function. However, the Log Gamma function computes more efficiently than do the ln and the Ln Gamma functions together.

squash(☐)

The **Squash** function is an efficient computation of the function

$$\frac{1}{1+e^x}$$

where *x* is any column, expression, or temporary variable.

Character Functions

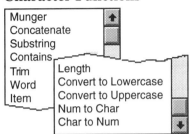

Character functions accept character arguments or return character strings and can convert the data type of a value from numeric to character, or from character to numeric.

Each of these functions are described here:

Munger

can compute new character strings from existing strings by inserting or deleting characters. It can also produce substrings, calculate indexes, and perform other tasks depending on how you specify its arguments. Munger's four arguments are the *search string,* the *offset,* the *find string,* and the *replace string* as shown in **Figure 5.7**.

Figure 5.7 Arguments of the Munger Function

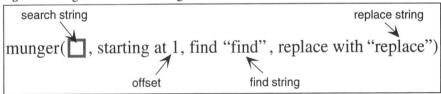

Note➜ The Munger function treats uppercase and lowercase letters as different characters.

Search String

is a character expression. Munger applies the other three arguments to this string to compute a result.

Offset or Starting Position

is a numeric expression. If the offset is greater than the search string's length, Munger uses the string's length as the offset.

Find Value

is a character or numeric expression. Use a character string as search criterion, or use a positive integer to represent the number of consecutive characters to find. If you specify a negative integer as the Find Value, Munger searches through to the end of the string.

Replace String

can be a string or a missing term. If it is a string, Munger replaces the search criterion with the Replace String to form the result. If it is a missing term, Munger calculates either a substring or an index depending on whether the Find Value is an integer or a string, respectively.

The following examples show uses of the Munger function. Assume a character column of names with "Katie Layman" as one of its values. To simplify the examples, the literal name "Katie Layman" is the search string instead of a column name.

Insert Characters

This example finds the blank between the first and last name, and inserts the middle initial "M.". The formula

    munger("Katie Layman", starting at 1, find " ", replace with " M. "   )

inserts the middle initial M., and evaluates as Katie M. Layman.

Delete Characters

To delete one or more characters start at the first position of the string, designate the characters to delete as the find string, and enter a replace string that has no characters. For example, the function,

    munger("Katie Layman", starting at 1, find "i", replace with ""   )

removes the "i" from Katie and evaluates as Kate Layman.

Note➡ A replace field with a null (no value) string is different from a replace field that is deleted altogether, If you delete the replace string, it appears as an empty box, as shown in the next Munger example.

Find the Position (Index) of a Value

When the replace field is not defined, Munger behaves like an index function and returns the numeric position of the first character of the search string, if it exists. For example

    munger("Katie Layman", starting at 1, find " ", replace with   ▢)

searches for a single blank and finds it in position 6. If you use the Munger function in this way and calculate a numeric value in a character column, an error dialog appears with a button to that lets you change the column type to numeric. If the search string is not found, Munger returns a zero. This use of Munger produces the same result as the Contains function, described later in this section.

Find a substring

Munger can extract substrings. For example, to see the first name only

    munger("Katie Layman", starting at 1, find 5, replace with   ▢)

starts at position 1, reads through position 5, and ignores the remaining characters because the replace function is not defined. This yields "Katie." This use of Munger produces the same result as the Substring, described later in this section.

Using Munger, an alternative way to find a substring, is with a start value, any negative find value, and a null replace value. This example also computes "Katie."

    munger("Katie Layman", starting at 6, find -1, replace with ""   )

To extract a substring, you can also use the Substring function described later in this section.

The following examples show additional character functions:

☐||☐ (concatenate character strings)

The Concatenate function (shown as two vertical bars) produces a string with the function's second character argument appended to the first. For example,

"Dr. "||*name*

produces a new string consisting of the title "Dr." followed by a space and the contents of the *name* string.

substring(☐, starting at ☐, of length ☐)

The **substring** function produces extracts the characters that are the portion of the first argument beginning at the position given by the second argument and ending with the value in the third argument. The first argument can be either a character column or a literal value. The starting argument and the length argument can be numbers of expressions that evaluate to numbers.

For example, to show the first name only,

substring("Katie Layman", starting at 1, of length 5)

starts at position 1, reads through position 5, and ignores the remaining characters, which yields "Katie."

☐ contains ☐

The **Contains** function returns the numeric position within the first argument of the second argument if it exists, and a zero otherwise. For example,

"Katie Layman " contains "L"   evaluates as 7

"Lillie Layman " contains "L"   evaluates as 1.

trim ☐

The **Trim** function produces a new character string from its argument, removing any trailing blanks. For example,

length "John   "   evaluates as "John".

word(1, of ☐, delimited by " ")
item(1, of ☐, delimited by ",")

The **Word** function and the **Item** function both extract a word from a character string. The delimiter you enter as the last argument defines what a word is, and the first argument tells the position of the word you want. For example, to extract the last name in the following examples, use a blank as the delimiting character and ask for the second word. The two following functions both return the word "Katie."

word(2, of "Katie Layman", delimited by " ")

item(2, of "Katie Layman", delimited by " ")

The item function is different than the Word function because of the way they treat word delimiters. If a delimiter is found multiple times or you enter a delimiter with multiple characters, the word function treats them as a single delimiter. The item function uses each delimiter to define a new word position.

To compare, suppose a name is of the form lastname, firstname. The word delimiter is a comma followed by a blank.

    word(2, of "Layman, Katie ", delimited by ", ")

    item(2, of "Layman, Katie ", delimited by ", ")

The Word function treats the comma and blank as a single delimiter and returns "Katie" as the second word. The item function returns a missing value because it treats the comma and blank separately and finds nothing between them.

length □

The Length function calculates the length of its argument. For example,

    length "Elizabeth"   evaluates as 9.

lowercase □
uppercase □

The Convert to Lowercase function converts any upper case character found in its argument to the equivalent lowercase character.

The Convert to Uppercase function converts any lowercase character found in its argument to the equivalent lower case character. For example,

    lowercase "KATIE LAYMAN" evaluates as "katie layman", and

    uppercase "katie layman" evaluates as "KATIE LAYMAN".

char □

The Num to Char function produces a character string that corresponds to the digits in its numeric argument. For example,

    char 1.123 evaluates as "1.123".

num □

The Char to Num function produces a numeric value that corresponds to its character string argument when the character string consists of numbers only. All other character data produce a missing value. For example,

    num "1.123"   evaluates as 1.123.

Comparisons

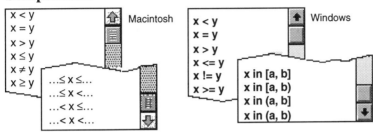

Comparisons are functions that compare the ranks of their arguments. Each comparison relationship evaluates as *true* or *false* based on numeric magnitudes or character rankings.

Comparisons are most useful when you include them in conditional expressions, but they can also stand alone as numeric expressions. A true relationship evaluates as 1, and a false relationship evaluates as 0.

The relational symbols are

- < (less than)
- > (greater than)
- ≥ (greater than or equal to)
- = (equal to)
- ≤ (less than or equal to)
- ≠ (not equal to)

A relational symbol's arguments can be any two expressions. However, both arguments in a Comparison function must be of the same data type. Each data type accepts a missing term. A missing term is considered less than all other values.

The calculator uses the International Utilities package when comparing character strings. This package contains different rankings for each international character set and takes diacritical marks into consideration.

Often, you can express two comparisons as one range comparison. A range comparison has three arguments and evaluates as true whenever the specified variable falls within the designated interval. For example, the expression

$$100 < Weight < 150$$

evaluates as 1 when a value in the weight column is between 100 and 150. This comparison is an efficient way of expressing

$$(100 < Weight) \text{ and } (Weight < 150)$$

The function browser shows relational symbols comparing x's to y's. To enter a comparison into a formula, highlight an expression in the formula and click the appropriate relationship from the comparison list. The comparison appears in the formula display, and your highlighted expression replaces the x in a comparison, or the lower bound (the first ellipsis) in a range comparison. Missing terms appear as place holders for the other arguments.

Conditions

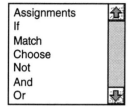

JMP offers four types of conditional expressions (called *conditionals* for short) in the Conditions list. They are **Assignments**, **If**, **Match**, and **Choose**. These expressions let you build a sequence of clauses paired with *result expressions*. Constructing a sequence of clauses is the way you *conditionally* assign values to cells in a calculated column.

Assignments is used with temporary variables (discussed later in this chapter) to simplify complicated expressions.

With **If** and **Match** the calculator searches down from the top of the sequence for the first true clause and evaluates the corresponding result expression. Subsequent true clauses are ignored. For maximum efficiency, list the most frequently evaluated clause/result pairs first in the sequence. When no clause is true, the calculator evaluates the result expression that accompanies the *otherwise* clause. With **Choose** the calculator goes directly to the correct choice clause and evaluates the result expression.

To build a conditional expression first select a type of conditional: **Assignments**, **If**, **Match**, or **Choose**. Then use the insert and delete clause buttons on the calculator panel to expand the expression. Each feature is described below.

Note→ All result expressions in a conditional expression must evaluate to the same data type. A missing term matches any data type. A missing term evaluates as a missing value of the appropriate type. By definition, expressions that evaluate as zero or missing are *false*. All other numeric expressions are *true*.

The following examples help show how to use conditionals:

Assignments

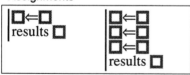

The Assignments function is used to assign expressions to temporary variables, which are then used in a complex equation. This technique can greatly simplify building an equation. For example, the formula for a Weibull loss function is

$$\begin{cases} \dfrac{Model}{sigma} - e^{\left(\dfrac{Model}{sigma}\right)} - \ln sigma, & \text{if } Censor \neq 0 \\ -e^{\left(\dfrac{Model}{sigma}\right)}, & \text{otherwise} \end{cases}$$

where **Model** is a data table column and **sigma** is a parameter.

You can simplify this complicated formula. Choose **Assignments** and use the insert button to create the assignment structure

$$\begin{vmatrix} \square \Leftarrow \square \\ \square \Leftarrow \square \\ \text{results } \square \end{vmatrix}$$

Suppose you create two temporary variables, *t1* and *t2* and use them as follows:

$$\begin{vmatrix} t1 \Leftarrow \dfrac{Model}{\textbf{sigma}} \\ t2 \Leftarrow e^{t1} \\ \text{results } \square \end{vmatrix}$$

The results expression is constructed by substituting *t1* and *t2* into the Weibull function shown previously, which gives

$$\begin{vmatrix} t1 \Leftarrow \dfrac{Model}{\textbf{sigma}} \\ t2 \Leftarrow e^{t1} \\ \text{results } - \begin{cases} t1-t2-\ln \textbf{sigma}, & \text{if } Censor \neq 0 \\ -t2, & \text{otherwise} \end{cases} \end{vmatrix}$$

The results expression is easier to manage and improves computational efficiency.

If

$$\begin{cases} \square, & \text{if } \square \\ \square, & \text{otherwise} \end{cases}$$

When you highlight an expression and click **If**, the calculator creates a new conditional expression with one **If** clause and one **otherwise** clause as shown above. The highlighted expression becomes the first result expression.

A conditional expression is usually a comparison. However, any expression that evaluates as a numeric value can be a conditional expression.

Note→ By definition, expressions that evaluate as zero or missing are *false*. All other numeric expressions are *true*.

To enter a conditional expression, fill the three missing terms with expressions. Use insert and to add new **If** clauses or remove unwanted clauses.

For example, enter the conditional expression

$$\begin{cases} \dfrac{count}{total} \cdot 100, & \text{if } total \neq 0 \\ 0, & \text{otherwise} \end{cases}$$

to calculate *count* as a percentage of *total* when *total* is not 0.

Match

$$\left\{\begin{array}{l}\text{match } \square: \\ \square, \quad \text{when } \square \\ \square, \quad \text{otherwise}\end{array}\right.$$

When you highlight an expression and click **Match**, the calculator creates a new conditional expression with one **when** clause and one **otherwise** clause as shown above. The highlighted expression becomes the **Match** argument.

To enter a conditional expression, fill the four missing terms with expressions. Use insert 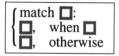 and delete to add a new when clause or remove unwanted clauses.

The **Match** conditional compares an expression to a list of clauses and returns the value of the result expression for the first matching clause encountered. With **Match**, you provide the matching expression only once and then give a match for each clause.

For example, to assign character class values to numeric fat to protein ratios, enter the conditional expression

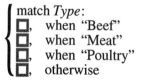

$$\left\{\begin{array}{ll}\text{match } \textit{Fat to Protein ratio} : \\ \text{"Very Lean"}, & \text{when } 0.2 \\ \text{"Lean"}, & \text{when } 0.25 \\ \text{"Average"}, & \text{when } 0.33 \\ \text{"Fatty"}, & \text{when } 0.5 \\ \text{"Very Fatty"}, & \text{when } 1 \\ \text{"Unknown"}, & \text{otherwise}\end{array}\right.$$

Note→ A shortcut for building **Match** conditionals with **when** clauses for all values of a nominal or ordinal variable occurring in the data is available with option–click. For example, suppose you want a **Match** conditional for the nominal variable **Type** from the HOTDOGS data (hotdogs.jmp in the Windows data folder) with values Beef, Meat, and Poultry. First choose the column **Type** from the column selector list. Then with **Type** selected in the formula display, option–click on **Match** in the function browser.

The formula that results is

$$\left\{\begin{array}{l}\text{match } \textit{Type}: \\ \square, \quad \text{when "Beef"} \\ \square, \quad \text{when "Meat"} \\ \square, \quad \text{when "Poultry"} \\ \square, \quad \text{otherwise}\end{array}\right.$$

Note→ **Match** evaluates faster and uses less memory than an equivalent **If**. See the section **Efficiency** in Chapter 6, "Using the Calculator," for a comparison of **Match** and **If** conditionals.

Choose

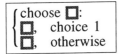

Choose is a special case of Match in which the arguments of when clauses are a sequence of integers starting at 1. Choice clauses replace when clauses in the Choose conditional.

When you highlight an expression and click Choose, the calculator creates a new conditional expression with one choice clause and one otherwise clause as shown above. The highlighted expression becomes the Choose argument.

To enter a conditional expression, fill the three missing terms with expressions. Use insert [▲] and delete [✐] to add new choice clauses or remove unwanted clauses.

The Choose conditional evaluates the choose expression and goes immediately to the corresponding result expression to generate the returned value. With Choose, you provide a choosing expression that yields sequential integers starting at 1 only once, then you give a choice for each integer in the sequence.

For example, to assign character class values to numeric sensory ratings between 1 and 4, enter the conditional expression

$$\begin{cases} \text{choose } Sensory\ Rating : \\ \text{"Excellent"}, \quad \text{choice 1} \\ \text{"Good"}, \quad \text{choice 2} \\ \text{"Fair"}, \quad \text{choice 3} \\ \text{"Poor"}, \quad \text{choice 4} \\ \text{"Unknown"}, \quad \text{otherwise} \end{cases}$$

The next examples show functions in the Conditions list that can be part of conditional expressions:

not ☐

When its argument is false, the Not function evaluates as 1. Otherwise, the Not function evaluates as 0.

When you apply the Not function, use parentheses where necessary to avoid ambiguity. For example, not $(weight = 64)$ can be either true or false (either 1 or 0), but $(\text{not } weight) = 64$ is always false (0) because Not can only return 0 or 1. Expressions such as not $(weight = 64)$ can also be entered as $weight \neq 64$.

▢ and ▢

The And function evaluates as 1 when both of its arguments are true. Otherwise it evaluates as 0 (see **Figure 5.8**).

The formula

$$\begin{cases} \text{"driver"}, & \text{if } \textit{licensed}=\text{"Yes"} \text{ and } \textit{age}>21 \\ \text{"passenger"}, & \text{otherwise} \end{cases}$$

labels participants as drivers only if both comparisons are true.

▢ or ▢

The Or function evaluates as 1 when either of its arguments is true. If both of its arguments are false, then the Or expression evaluates as 0 (see **Figure 5.8**).

The formula

$$\begin{cases} \text{"cabin 1"}, & \text{if } \textit{instructor}=\text{"Yes"} \text{ or } \textit{age}>21 \\ \text{"cabin 2"}, & \text{otherwise} \end{cases}$$

assigns instructors and all participants over 21 to cabin 1.

Figure 5.8 Evaluations of And and Or Expressions

and	True	False
True	1	0
False	0	0

or	True	False
True	1	1
False	1	0

Random Number Functions

Random number functions generate real numbers by effectively "rolling the dice" within the constraints of the specified distribution. The random number functions in JMP appear in formulas preceded by a "?" to indicate randomness. Each time you click **Evaluate** in the calculator window, these functions produce a new set of random numbers. This section describes the three random functions Uniform, Normal, and Shuffle.

?uniform

Uniform generates random numbers uniformly between 0 and 1. This means that any number between 0 and 1 is as likely to be generated as any other. The result is an approximately even distribution.

?normal

Normal generates random numbers that approximate a normal distribution with a mean of 0 and variance of 1. The normal distribution is bell shaped and symmetrical.. You can modify the Normal function with constants to specify a normal distribution with a different mean and standard deviation. For example, the formula

?normal •5+30

generates a random normal variable with a mean of 30 and a standard deviation of 5.

?shuffle

Shuffle selects a row number at random from the current data table. Each row number is selected only once. When Shuffle is used as a subscript, it returns a value selected at random from the column that serves as its argument. Each value from the original column is assigned only once as Shuffle's result. For example, to identify a 50% random sample without replacement, use the following formula:

$$\begin{cases} X_{?shuffle}, & \text{if } i \leq \frac{n}{2} \\ \square, & \text{otherwise} \end{cases}$$

This formula chooses half the values (n/2) from the column X and assigns them to the first half of the rows in the computed column. The remaining rows of the computed column fill with missing values.

Probability Functions

Normal Distribution
Normal Quantile
Chi-Squared Distribution
Chi-Squared Quantile
Student's t Distribution
Student's t Quantile
F Distribution
F Quantile

Probability functions calculate quantiles and probabilities for normal, chi–square, Student's t, and F distributions.

This section gives you examples of the probability functions.

normDist (☐)

The Normal Distribution function accepts a quantile argument. It returns the probability that an observation from the standard normal distribution is less than or equal to the specified quantile. For example, the expression

    normDist (1.96)

returns the probability that an observation from the standard normal distribution is less than or equal to 1.96. The expression evaluates as 0.975.

The Normal Quantile function is the inverse of the Normal Distribution function.

normQuant (☐)

The Normal Quantile function accepts a probability argument p. It returns the pth quantile from the standard normal distribution. For example, the expression

    normQuant (0.975)

returns the 97.5% quantile from the standard normal distribution. The expression evaluates as 1.96.

The Normal Distribution function is the inverse of the Normal Quantile function.

chi-square Dist (☐, ☐ DF, centered at 0)

The Chi–squared Distribution function accepts three arguments: a quantile, a degrees–of–freedom, and a noncentrality parameter. It returns the probability that an observation from the chi–square distribution with the specified noncentrality parameter and degrees of freedom is less than or equal to the given quantile. For example, the expression

    chi-square Dist (11.264, 5 DF, centered at 0)

returns the probability that an observation from the chi–square distribution centered at 0 with 5 degrees of freedom is less than or equal to 11.264. The expression evaluates as 0.95361.

The Chi–squared Distribution function accepts integer and noninteger degrees of freedom. It is centered at 0 by default. The Chi–squared Quantile function is the inverse of the Chi–squared Distribution function.

chi-square Quant (☐, ☐ DF, centered at 0)

The Chi–squared Quantile function accepts three arguments: a probability p, a degrees of freedom, and a noncentrality parameter. It returns the pth quantile from the chi–square distribution with the specified noncentrality parameter and degrees of freedom. For example, the expression

$\qquad$ chi-square Quant (0.95, 3.5 DF, centered at 4.5)

returns the 95% quantile from the chi–square distribution centered at 4.5 with 3.5 degrees of freedom. The expression evaluates as 8.665122.

The Chi–squared Quantile function accepts integer and noninteger degrees of freedom. It is centered at 0 by default. The Chi–squared Distribution function is the inverse of the Chi–squared Quantile function.

t Dist (☐, ☐ DF, centered at 0)

The Student's t Distribution function accepts three arguments: a quantile, a degrees of freedom, and a noncentrality parameter. It returns the probability that an observation from the Student's t distribution with the specified noncentrality parameter and degrees of freedom is less than or equal to the given quantile. For example, the expression

$\qquad$ t Dist (0.9, 5 DF, centered at 0)

returns the probability that an observation from the Student's t distribution centered at 0 with 5 degrees of freedom is less than or equal to 0.9. The expression evaluates as 0.79531.

The Student's t Distribution function accepts integer and noninteger degrees of freedom. It is centered at 0 by default. The Student's t Quantile function is the inverse of the Student's t Distribution function.

t Quant (☐, ☐ DF, centered at 0)

The Student's t Quantile function accepts three arguments: a probability p, a degrees of freedom, and a noncentrality parameter. It returns the pth quantile from the Student's t distribution with the specified noncentrality parameter and degrees of freedom. For example, the expression

$\qquad$ t Quant (0.95, 2.5 DF, centered at 0)

returns the 95% quantile from the Student's t distribution centered at 0 with 2.5 degrees of freedom. The expression evaluates as 2.558219.

The Student's t Quantile function accepts integer and noninteger degrees of freedom. It is centered at 0 by default. The Student's t Distribution function is the inverse of the Student's t Quantile function.

F Dist $\left(\square, \dfrac{1}{\square} \text{ DF, centered at } 0\right)$

The F Distribution function accepts four arguments: a quantile, a numerator and denominator degrees of freedom, and a noncentrality parameter. It returns the probability that an observation from the F distribution with the specified noncentrality parameter and degrees of freedom is less than or equal to the given quantile. For example, the expression

$$\text{F Dist}\left(3.32, \dfrac{2}{3} \text{ DF, centered at } 0\right)$$

returns the probability that an observation from the central F distribution with 2 degrees of freedom in the numerator and 3 degrees of freedom in the denominator is less than or equal to 3.32. The expression evaluates as 0.82639.

The F Distribution function accepts integer and noninteger degrees of freedom. By default it is centered at 0 and has 1 numerator degree of freedom. The F Quantile function is the inverse of the F Distribution function.

F Quant $\left(\square, \dfrac{1}{\square} \text{ DF, centered at } 0\right)$

The F Quantile function accepts four arguments: a probability p, a numerator and denominator degrees of freedom, and a noncentrality parameter. It returns the pth quantile from the F distribution with the specified noncentrality parameter and degrees of freedom. For example, the expression

$$\text{F Quant}\left(0.95, \dfrac{2}{10} \text{ DF, centered at } 3.2\right)$$

returns the 95% quantile from the F distribution centered at 3.2 with 2 degrees of freedom in the numerator and 10 degrees of freedom in the denominator. The expression evaluates as 9.520084.

The F Quantile function accepts integer and noninteger degrees of freedom. By default it is centered at 0 and has 1 numerator degree of freedom. The F Distribution function is the inverse of the F Quantile function.

Statistical Functions

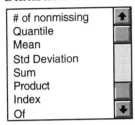

The calculator evaluates statistical functions differently from other functions. Most functions evaluate data for the current row only. However, all statistical functions require a set of values upon which to operate.

Note➡ The Sum and Product functions always evaluate for an explicit range of column values. All other statistical functions *always* evaluate for $i = 1$ to n values *on every row*.

Except for the # of non–missing function, statistical functions apply only to numeric data. The calculator excludes missing numeric values from its statistical calculations.

Each statistical function is described as follows:

N □

The # of non–missing function counts the number of *non*–missing values in the column you specify. A *missing numeric value* occurs when a cell has no assigned value or as the result of an invalid operation (such as division by zero). Missing values show on the spreadsheet as a missing value mark (•), illustrated in **Figure 5.9**. Missing character values are null character strings.

In formulas for row state columns, an excluded row state characteristic is treated as a missing value. The calculator interprets other missing values according to their data types.

Figure 5.9 Spreadsheet Representation of Missing Values

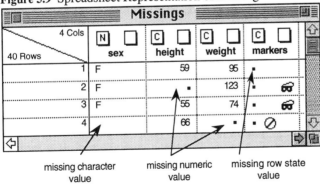

quantile$_{0.5}$ □

The Quantile function computes the value at which a specific percentage of the values is less than or equal to that value. For example, the value calculated as the 50% quantile, also called the *median*, is greater than or equal to 50% of the data. Half of the data values are less than the 50*th* quantile.

The Quantile function's subscript, *p,* represents the quantile percentage divided by 100. The 25% quantile, also called the lower quartile, corresponds to $p=.25$, and the 75% quantile, called the upper quartile, corresponds to $p=.75$. The default value of *p* is 0.5 (the median).

The calculator computes a quantile for a column of N nonmissing values by arranging the values in ascending order. The subscripts of the sorted column values, $y_1, y_2, ..., y_N$, represent the ranks in ascending order.

The *p*th quantile value is calculated using the formula $I = p \cdot (N+1)$ where *p* is the quantile and N is the total number of nonmissing values. If I is an integer, then the quantile value is $y_p = y_i$. If I is not an integer, then the value is interpolated by assigning the integer part of the result to i, and the fractional part to *f,* and by applying the formula

$$q_p = (1-f)y_i + (f)y_{i+1}$$

For example, suppose a column has values 2, 4, 6, 8, 10, 12, 14, 16, 18, and 20. The 50% quantile is calculated as follows:

.5 * (10+1) = 5.5

Because the result is fractional, the 50% quantile value is interpolated as

(1−.5) * 12 + (.5) * 10 = (.5)12 + (.5)10 = 6 + 5 = 11

The following are example Quantile formulas:

- quantile$_1$ *age* calculates the maximum *age*.
- quantile$_{0.75}$ *age* calculates the upper quartile *age*.
- quantile$_{0.5}$ *age* calculates the median *age*.
- quantile$_{0.25}$ *age* calculates the lower quartile *age*.
- quantile$_0$ *age* calculates the minimum *age*.

The formula
$$\text{quantile}_{\left(\frac{i-1}{n-1}\right)} age$$
lists the values of age in ascending order in the calculated column.

The formula
$$\text{quantile}_{\left(\frac{n-i}{n-1}\right)} age$$
lists the values of age in descending order in the calculated column.

▫

The Mean function calculates the mean (or arithmetic average) of the numeric values identified by its argument. The mean of a variable, y, is denoted by $\bar{y}$ and computed

$$\bar{y} = \frac{\sum_{i=1}^{n} y_i}{N}$$

The formula, $\overline{age}$, calculates the average of all nonmissing values in the *age* column.

std ▫

The Std. Deviation function measures the spread around the mean of the distribution identified by its argument. In the normal distribution, about 68% of the distribution is within one standard deviation of the mean, 95% of the distribution is within two standard deviations of the mean, and 99 percent of the distribution is within three standard deviations of the mean. The standard deviation of a variable, y, is usually denoted by *s* and computed

$$s = \sqrt{s^2}, \text{ where } s^2 = \frac{\sum_{i=1}^{N} (y-\bar{y})^2}{N-1}.$$

The formula, std *height*, calculates the standard deviation of the non–missing values in the column height.

$$\sum_{j=1}^{n} \square$$

The Sum function uses the summation notation shown above. To calculate a sum, replace the missing term with an expression containing the index variable j. Sum repeatedly evaluates the expression for j = 1, j = 2, through j = n and then adds the nonmissing results together to determine the final result.

You can replace n, the number of rows in the active spreadsheet, and the index constant, 1, with any expression appropriate for your formula.

For example, the following expression

$$\sum_{j=1}^{i} revenue_j$$

computes the total of all revenue values for row 1 through the current row number, filling the calculated column with the cumulative totals of the *revenue* column.

Note➔ See Index for an explanation of nesting summations.

$$\prod_{j=1}^{n} \square$$

The Product function uses the notation shown above. To calculate a product, replace the missing term with an expression containing the index variable j. Product repeatedly evaluates the expression for j = 1, j = 2, through j = n and multiplies the nonmissing results together to determine the final result.

You can replace n, the number of rows in the active spreadsheet and the index constant, 1, with any expression appropriate for your formula.

For example, the expression

$$\prod_{j=1}^{i} j$$

calculates *i*! (each row number's factorial).

Note➔ See Index next for an explanation of nesting products.

Index

Index works with Sum and Product, usually as a column subscript (see Terms). The calculator matches a new index to the first summation or product index variable it finds while searching outward from the highlighted expression. Each time you choose Index while one expression remains highlighted, the calculator advances it to the next index out.

Sometimes, it is useful to designate an expression containing a summation or product as the argument of a summation or product function. This is called *nesting*. The calculator automatically assigns a unique index variable for each Σ and Π, assigning j as the left–most index and reassigning the existing indexes as necessary. All corresponding index variables within the formula also adjust as necessary.

The expression

$$\sum_{j=1}^{n} \sum_{k=1}^{n} \blacksquare$$

becomes the following expression when index is applied.

$$\sum_{j=1}^{n} \sum_{k=1}^{n} \boxed{k}$$

Apply index again while k is highlighted to change k to j.

$$\sum_{j=1}^{n} \sum_{k=1}^{n} \boxed{j}$$

Because there are no further index variables, applying index again changes j back to the first variable, k.

Note➔ When columns are arguments of a summation or a product function they are automatically assigned the index of the innermost Σ or Π. The Index function can be applied to the column subscript to choose an index variable at another nesting level as described above.

of(☐)

The Of function is usually used with any of the statistical functions, which enables you to compute statistics across columns. For each row,

> std of(*hist0, hist1, hist3, hist5*)

computes the standard deviation of the variables in the argument list.

 To create a list of arguments for the Of operator, use either the COMMA key or the insert clause button on the calculator panel. Use the DELETE key or the delete clause button to remove unwanted arguments.

Date & Time Functions

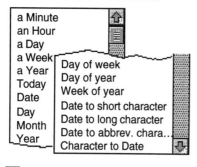

JMP stores dates and times in numeric columns, using the Macintosh standard of the *number of seconds since Jan 1, 1904*. When a column has date values, you can assign a date format to that column using the **Date & Time** format pop-up menu in the Column Info dialog so that they display in a familiar form. See Chapter 3, "JMP Data Tables," for more details about using dates.

The calculator supports JMP dates with the following functions:

- ☐ minutes
- ☐ hours
- ☐ days
- ☐ weeks
- ☐ years

Each of the functions listed above converts from the units of the function name to the equivalent number of seconds. The argument must be a number or numeric expression. For example,

2 minutes yields 120

1 year yields 31,557,600 (60 sec • 60 min • 24 hrs • 365.25 days)

today

The Today function returns the number of seconds between January 1, 1904 and the current date. For example, at midnight on March 20, 1991 (a Wednesday), the Today function returns

2752272000

(2,752,272,000 seconds) and continues counting. If you evaluate the Today function later in the day, it reflects the additional seconds.

date(day ☐, month ☐, year ☐)

The Date function accepts numeric expressions for day, month, and year and returns the associated JMP date. For example,

date(day 20, month 3, year 1991)

always evaluates to 2752272000.

day(☐)
month(☐)
year(☐)

The argument for the Day, Month, and Year functions is interpreted as a JMP date. These functions returns the day of the month, the month as a number from 1 to 12, and a four–digit year, respectively

For example, on March 20, 1991, both

day(today) and day(2752272000) return the number 20,

month(today) and month(2752272000) return the number 3, and

year(today) and year(2752272000) return the number 1991.

dayOfWeek(☐)
dayOfYear(☐)
weekOfYear(☐)

The argument for Day of week, Day of year, and Week of year functions is a JMP date.

Day of week returns a number from 1 to 7 where 1 represents Sunday, Day of year returns the number of days from the beginning of the year, and Week of year returns a number from 1 to 52.

For example, on Wednesday, March 20, 1991, both

dayOfWeek(today) and dayOfWeek(2752272000) return the number 4,

dayOfYear(today) and dayOfYear(2752272000) return the number 79, and

weekOfYear(today) and weekOfYear(2752272000) return the number 12.

char (☐ as short date)
char (☐ as long date)
char (☐ as abbrev. date)

The argument for the date as *format* character functions is a JMP date. They create character strings that are the formatted representation of the argument.

For example,

char (2752272000 as short date) returns "3/20/91"

char (2752272000 as long date) returns "Wednesday, March 20, 1991"

char (2752272000 as abbrev. date) returns "Wed, Mar 20,1991".

num (☐ as date)

The argument for the Character to Date function is a date character string, such as "Wednesday, March 20, 1991," or any valid string recognized by your machine as a date. This function returns the appropriate JMP date value.

Row State Functions

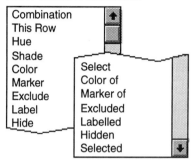

Formulas process row state data just as they process character and numeric data. There are six row state conditions called Select, Hide, Exclude, Label, Color, and Marker:

- Select rows to identify data applicable to JMP commands.
- Hide rows to hide points in report displays.
- Exclude rows to eliminate data from analysis calculations and displays.
- Label rows to identify points in plots.
- Use one or more of the 64 JMP colors to distinguish points in report displays.
- Use one or more of the 8 JMP markers to distinguish points in report displays.

A row can be assigned any combination of row states. A list of multiple row states is called a *row state combination*. Row state functions are described as follows:

«☐, ☐»

New Combination generates a row state combination with two arguments. The currently selected expression becomes the first argument when you choose New Combination. Replace each argument with an expression that evaluates to a row state.

The formula

«select (i mod 2), label ((i+1) mod 2)»

alternately labels or selects each row in the calculated row state column. The select and label functions are defined later in this section.

Use the insert and delete buttons in the calculator panel to add more arguments or remove unwanted arguments.

Note→ If you include conflicting row states in a combination, the results are unpredictable.

row

The This row function, denoted *row* in formulas, returns the active row state condition of the current row as true or false. You can use this function to conveniently write conditional clauses that depend on the status of the current row. For example,

$$\begin{cases} 1, & \text{if selected row and labelled row} \\ 0, & \text{otherwise} \end{cases}$$

assigns a 1 to rows that are currently selected and labeled and a zero otherwise.

hue ☐

The Hue function returns the color from the JMP Hue Map that corresponds to its integer argument. JMP hues are numbered 0 through 11 but larger integers are treated as *modulo 12*. The Hue function does not map to black, gray, or white. A Hue of 0 maps to red and Hue of 11 maps to magenta. The formula

$$\text{hue}\left(\frac{z-\text{quantile}_0 z}{\text{quantile}_1 z - \text{quantile}_0 z} \cdot 12\right)$$

assigns row state colors in a chromatic spread based on the value of z. If you hold down the OPTION key (ALT key under Windows) and choose Hue, the following formula appears:

$$\text{hue}\left(\begin{vmatrix} \mathbf{\textit{min1}} \Leftarrow \text{quantile}_0 z \\ \mathbf{\textit{max1}} \Leftarrow \text{quantile}_1 z \\ \text{results } \dfrac{z - \mathbf{\textit{min1}}}{\mathbf{\textit{max1}} - \mathbf{\textit{min1}}} \cdot 12 \end{vmatrix}\right)$$

This formula assigns row state colors in the same way as the first formula but creates special variables *minl* and *maxl* to form a simplified results formula. Special variables are discussed later in this chapter.

shade ☐

The Shade function assigns 5 shade levels to a color or hue. A shade of -2 is darkest and shade of +2 is lightest. Shade of zero is a pure color. If you hold down the option key and choose Shade, the following formula appears:

$$\text{shade}\left(\begin{vmatrix} \mathbf{\textit{min2}} \Leftarrow \text{quantile}_0 z \\ \mathbf{\textit{max2}} \Leftarrow \text{quantile}_1 z \\ \text{results } \dfrac{z - \mathbf{\textit{min2}}}{\mathbf{\textit{max2}} - \mathbf{\textit{min2}}} \cdot 5 - 2 \end{vmatrix}\right)$$

This formula assigns shade values based on the value of z. However, to assign all shades of all the colors in the colors palette, you need to use the Hue and Shade assignments together as follows:

$$\Biggl\langle\!\!\Biggl\langle \text{hue}\left(\begin{vmatrix} \mathbf{\textit{min2}} \Leftarrow \text{quantile}_0 z \\ \mathbf{\textit{max2}} \Leftarrow \text{quantile}_1 z \\ \text{results } \dfrac{z - \mathbf{\textit{min2}}}{\mathbf{\textit{max2}} - \mathbf{\textit{min2}}} \cdot 12 \end{vmatrix}\right), \text{shade}\left(\begin{vmatrix} \mathbf{\textit{min1}} \Leftarrow \text{quantile}_0 z \\ \mathbf{\textit{max1}} \Leftarrow \text{quantile}_1 z \\ \text{results } \dfrac{z - \mathbf{\textit{min1}}}{\mathbf{\textit{max1}} - \mathbf{\textit{min1}}} \cdot 5 - 2 \end{vmatrix}\right)\Biggr\rangle\!\!\Biggr\rangle$$

The formula above uses the Combination function described at the beginning of this section. The first argument in the Combination function is OPTION–Hue (or ALT–Hue) for the variable x, and the second argument is OPTION–Shade (or ALT–Shade) for y.

color ▢

The Color function returns the color from the JMP Color Map that corresponds to its integer argument. JMP colors are numbered 0 through 12, but larger integers map to color indices are treated as *modulo12*. Zero map to black.

marker ▢

The Marker function returns markers from the JMP Marker Map that correspond to its integer argument. JMP markers are numbered 0 through 7, but larger integers that map to marker indices are treated as *modulo 8*.

The formula

marker i

assigns all the row state markers in a repeating sequence to the calculated row state column.

Exclude, Label, Hide, and Select

The Exclude, Label, Hide, and Select functions accept expressions that evaluate as either 1 or 0 (true or false) as an argument. Enter 1 to activate one of these row states or 0 to deactivate a row state.

For example, exclude i mod 2 excludes all odd–numbered rows in the calculated row state column.

colorOf ▢

Color of accepts a row state argument and returns a number from the JMP Color Map that corresponds to the color argument. Use the row state pop–up menu on the calculator panel to select a color argument for colorOf.

markerOf ▢

Marker of accepts a row state argument and returns a number from the JMP Marker Map that corresponds to the marker argument.

For example, markerOf ◆ is 4.

Use the row state pop–up menu on the calculator panel to select a color argument for markerOf.

Excluded, Labeled, Hidden, and Selected

The Excluded, Labeled, Hidden, and Selected functions accept any row state expression and return a 1 if the row state is active or 0 if the row state is inactive. These characteristics are inactive by default.

The formula

exclude (excluded *Status* or hidden *Status*)

excludes all rows if they are excluded or hidden in a row state column named *status*.

Parameters

Parameters are named constants created in the calculator that can be used in any formula. Numeric parameters are most useful in formulas created for nonlinear fitting.

Clicking the New Parameter item at the top of the list brings up the dialog shown in **Figure 5.10**. You use the dialog to assign a name, value, and data type to the new parameter. The default names for parameters are p1, p2, and so on. You can create as many parameters as you need, which are then available to all columns that have formulas. After a parameter is created, it shows at the bottom of the function browser parameter list.

Figure 5..10 The New Parameter Dialog

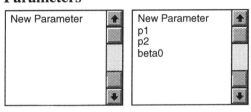

You assign a data type (numeric, character, or row state) with the **Data Type** pop–up menu. Click **Done** when the dialog is complete or **Cancel** to exit the dialog without creating a new parameter. To change parameter settings or remove a parameter, OPTION–click (ALT–click under Windows) the parameter name. This displays the dialog in **Figure 5.10** again, with the **Remove** button activated.

Parameters are added to formulas in much the same way that variables are added. To insert parameters into your formula, click the parameter name in the function browser list. Parameters are easy to recognize in formulas because they are displayed in **bold** type. For example, in the formula

$$\mathbf{b0} \cdot \left(1 - e^{-\mathbf{b1} \cdot X}\right)$$

b0 and b1 are parameters.

Parameter data types must be valid in the context of the expressions where they are used. If a parameter data type is invalid, an error message appears. When a parameter is used in a model for the nonlinear platform, the initial value or starting value should be given as the parameter value. After completing a nonlinear fit or after using the **Reset** button in the nonlinear control panel, the parameter's value is the most recent value computed by the nonlinear platform.

When you paste a formula with parameters into a column, the parameters are automatically created for that column unless it has existing parameters with the same names.

Variables

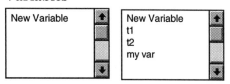

The Variables function lets you define temporary variables to use in expressions. Temporary variables exist only for the column in which they are defined.

Clicking New Variable brings up the dialog shown in **Figure 5.11**. You use the dialog to assign a name and data type (**Numeric, Character,** or **Row State**) to the new variable. By default, temporary variables are named t1, t2, and so on.

Figure 5.11 The Define New Variable Dialog

Temporary variables are most often used with the Assignments function, previously described in the Conditions function section. Assignments is used to assign expressions to temporary variables, which are then used in a complex equation. This technique can greatly simplify building an equation and can improve the efficiency of its evaluation.

For example, if Model is a data table column and **Sigma** is a parameter, the formula for a Weibull loss function is

$$\begin{cases} \dfrac{Model}{\text{sigma}} - e^{\left(\frac{Model}{\text{sigma}}\right)} - \ln \text{sigma}, & \text{if } Censor \neq 0 \\ -e^{\left(\frac{Model}{\text{sigma}}\right)}, & \text{otherwise} \end{cases}$$

This complicated formula can be simplified. Choose Assignments from the Conditions function list and use the insert button on the calculator panel to create the assignment structure shown to the left in **Figure 5.12**. Then create two temporary variables, *t1* and *t2* , and use them as shown in second assignment function. The results expression is constructed by substituting *t1* and *t2* into the Weibull function, which gives the results expression shown to the right in **Figure 5.12**.

Figure 5.12 Using Temporary Variables

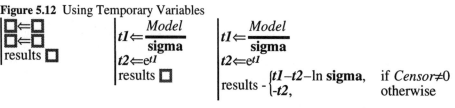

Editing Functions

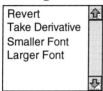

Editing features can help you build formulas with ease. Each feature is described as follows:

Revert

discards all changes made to the formula after the data table was last saved.

Take Derivative

takes the derivative of a formula with respect to any highlighted column or parameter within the formula. The derivative replaces the original formula in the formula display.

To take a derivative, first creat a formula. Next, select any occurrence of a column or parameter within the formula. Click **Take Derivative** to replace the original formula by its derivative with respect to the selected term.

For example, the derivative of

$$b0 \cdot (1 - e^{-b1 \cdot x})$$

with respect to b1 shows as

$$b0 \cdot x \cdot e^{-b1 \cdot x}.$$

Use the Undo command to revert to the original formula.

Smaller Font

decreases the type sizes in the formula display. Using **Smaller Font**, you can view a large formula without scrolling. You can also use **Smaller Font** to adjust a formula's size before copying it.

The formula

$$10.75 + \left(\left(\begin{cases} 0.525, & \text{if } \textit{oil amt} = \text{``lots''} \\ -0.525, & \text{if } \textit{oil amt} = \text{``little''} \\ \bullet, & \text{otherwise} \end{cases} \right) + \left(\begin{cases} 1.75, & \text{if } \textit{batch} = \text{``small''} \\ -1.75, & \text{if } \textit{batch} = \text{``large''} \\ \bullet, & \text{otherwise} \end{cases} \right) \right)$$

has been reduced from the default 18 points to 10 points to fit into the space above.

Larger Font

increases the type sizes in the formula display. Using **Larger Font**, you can get a better view of the formula. You can also use **Larger Font** to adjust a formula's size before copying it.

The formula display in **Figure 5.13** shows part of a formula enlarged for a better view. To see more of any large formula, stretch the calculator window using the size box in the lower right corner of the window or scroll through the formula using the scroll bars.

Figure 5.13 Enlarged View in Formula Display

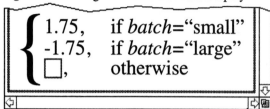

Note➡ The initial type style and size of characters in the formula display depend on the selection of fonts available in your machine's operating system. On the Macintosh the calculator first searches for the Times font and then the New York font. If neither font is available, formulas display in your system's default application font (usually Geneva). The calculator automatically chooses the largest type size available that is smaller than 24 point.

Chapter 6
Using the Calculator

The mathematical functions and other calculator tools discussed in the previous chapter give you a powerful and flexible way to create and manage data. This chapter tells you more about how to build JMP calculator formulas. The section on caution and alert messages tells you how to respond when you and the calculator have a communication failure.

The section on table templates is an introduction to a powerful way to use JMP tables stored with formulas tailored for specialized work such as

- constructing grids for plotting
- simulating sampling distributions
- nonlinear modeling.

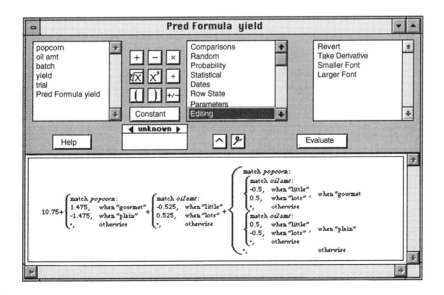

Chapter 6
Contents

Working with Formulas 179
   Computational Order of Precedence..179
   Building a Formula in Order of Precedence...179
   Constant Expressions 180
   Focused Work Areas..180
   Cutting and Pasting Formulas 181
   Selecting Expressions..181
   Dragging Expressions...182
   Changing a Formula...183
   Efficiency...184
Caution and Error Messages 185
   Caution Alerts..185
   Stop Alerts..187
Table Templates..188
   Grid Coordinates...189
   Actuarial Life Tables 189
   The Central Limit Theorem 190
   Nonlinear Modeling..190
Keyboard Shortcuts...191

Working with Formulas

Computational Order of Precedence

As you build a formula, keep in mind that all functions have an order of precedence defined by levels 1 through 6 as described in **Table 6.1**, where 1 is the highest order of precedence. Expressions with a high order of precedence are evaluated before those at lower levels. When an expression has operators of equal precedence, it is evaluated from left to right. You can use parentheses to override other precedence rules when necessary because any expression within parentheses is always evaluated first.

Terms have no order of precedence because they cannot be evaluated further.

Table 6.1 Precedence Order of Functions

Level 1:	Parentheses
Level 2:	Functions: |x|[†], Ceiling[†], Char to Num, Color, ColorOf, Log, Conditions[†], Exclude, Floor[†], Hide, Hidden, Label, Labeled, Length, Marker, MarkerOf, Max[†], Min[†], Mean[†], Munger, N, In, Num to Char, Power, Probability[†], Product, Quantile, Root, Row State Combinations[†], Select, Selected, Std. Deviation, Summation, Transcendental[†], Trim
Level 3:	*, ÷, Modulo
Level 4:	+, −
Level 5:	Comparisons: <, ≤, =, ≠, >, ≥, ≤ x ≤, ≤ x <, < x ≤, < x <
Level 6:	Logical Functions: And, Not, Or
†	When one of these functions has an expression as its argument, the argument has a higher order of precedence as it would if enclosed in parentheses.

Building a Formula in Order of Precedence

It is best to build a formula starting with any expression that serves as an argument. This is because functions have a high order of precedence and are always grouped with their corresponding arguments. It is a good idea to create expressions working from highest to lowest order of precedence when possible. If you need parentheses, be sure to click the left parenthesis in the keypad before entering the expression to be enclosed.

For example, given a data table with the columns A, B, and C, use the following steps to compose the expression A•(B+C). Note that this expression is not the same as A•B+C, which evaluates as (A•B)+C.

1. While the first missing term is highlighted ■, click column A in the column selector list.

2. Click the multiplication button in the keypad.

   A • ■

3. While the new missing term is highlighted, click the left parenthesis in the keypad to apply parentheses.

   A • (■)

4. Click column B in the column selector list.

5. While column B is still highlighted, click the addition button in the keypad.

6. While the new missing term is highlighted, click column C in the column selector list.

Because order of precedence determines which arguments are affected by each function, order of precedence also affects the grouping of expressions. Highlight functions in the formula to verify how the order of precedence rules have been applied.

Constant Expressions

Once JMP has evaluated a formula, you can highlight a constant expression to see its value in the constant entry field. This is true for both parameters and expressions that evaluate to a constant value.

Focused Work Areas

Functions, column names, and constants can only be entered into formulas when their area of the calculator is *focused*. Clicking the mouse in the column selector list, function browser, constant entry field, or formula display area, causes a keyboard focus shift to that item. For example, clicking a column name in the column selector list makes the scroll bars visible and the column name highlight. Likewise, clicking any function group in the function browser, makes the scroll bars of both function browser panels appear and the first function of the selected group highlight.

When the formula display area is focused, the gray border around it changes to black. Keystrokes and editing commands now affect the formula.

There are several methods of changing focus. One way is to use the mouse to make all selections. You can also repeatedly press the tab key on the keyboard to change focus by cycling through the column selector list, constant entry field, function browser, and formula display area. Also, typing the name of a column changes focus to the column selector list and selects the first column beginning with the typed letters. Typing a digit changes the focus to the constant entry field with a **numeric** data type. Typing a quote changes the focus to the constant entry field with a **character** data type.

Cutting and Pasting Formulas

You can cut or copy an entire formula or any expression and paste it into another formula display. If you paste a formula in the scrapbook, it is saved both as PICT and Expr (expression format) data. You can copy expression format data into any formula display, and you can paste PICT format information into other Macintosh applications.

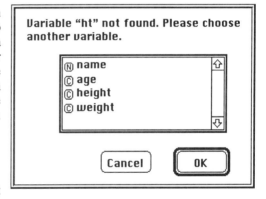

When you copy an expression from one data table to another, it expects to find the same column names. If a formula column name does not appear in the table, a dialog like the one shown to the right asks you for a substitute from the list of available column names. For example, you could change the formula

$$\frac{ht}{wt} \quad \text{to} \quad \frac{height}{weight}$$

by choosing height for Ht and weight for Wt.

Selecting Expressions

You can use the keyboard arrow keys select expressions for editing or to view the order of precedence within a formula when parentheses are not present. Clicking an operand in an associative expression, such as addition or multiplication, selects the operand. Clicking any associative operator, such as a + or • sign, selects the operator and its operands.

Once an operand is selected, the left and right arrow keys move the selection across other associative operands having equal precedence within the expression. The up arrow extends the current selection by adding the operand and operator of higher precedence to the selection. The down arrow reduces the current selection by removing an operand and operator from the selection.

The following example demonstrates how the arrow keys navigate through expressions:

Figure 6.1 Illustration of Selecting Expressions

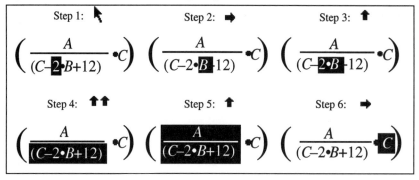

1. Select a single term in the denominator with the mouse.
2. Press the right arrow key once to select the second operand, *B*, of the • operator.
3. To select the entire • operation associated with *B*, press the up arrow key (note that clicking the mouse on the • operator results in the same selection). At this point the left and right arrows select the *C* and 12, respectively. The expression 2•*B* acts as an operand in both the + and − operations because addition and subtraction have a lower precedence than multiplication.
4. Select the denominator expression excluding the parentheses by pressing the up arrow twice. The first up arrow extends the selection to include the − operation and *C* operand. The second up arrow adds the + operation and 12 operand. Pressing the up arrow one more time extends the selection to include the parentheses. When the entire denominator is highlighted, the left and right arrows toggle the selection between the numerator and denominator.
5. Applying the up arrow with the denominator (or numerator) highlighted selects the left operand of the outermost • operation. Using the up arrow once more selects the complete • operation; twice, usint the up arrow twice selects the outermost parentheses and everything they enclose.
6. Finally, applying the right arrow applied to the left operand of the • operation highlights the right operand *C*.

Dragging Expressions

When you place the arrow cursor inside a selected expression, it changes to a hand cursor. This enables you to drag the selected expression to a new location. As you drag the expression, possible destinations highlight. The originally selected expression replaces the highlighted expression when the mouse button is released.

The following steps show how to move the text "Male" from the female to the male result clause in the sex match statement in **Figure 6.2**:

1. Press the mouse button over the selected term "Male".
2. Without releasing the mouse, move the hand until the male result clause highlights.
3. Release the mouse button and the text "Male" moves from the female to the male result clause.

Figure 6.2 Example of Dragging Expressions

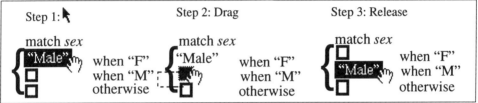

Using the OPTION key (ALT key under Windows) with the hand enables you to drag a copy of an expression to a new destination. An example of OPTION–dragging (or ALT–dragging) is shown in **Figure 6.3**. Using the COMMAND key on the Macintosh or the CONTROL key under Windows allows you to swap the selected expression with the destination expression. To swap or drag a copy of an expression you must press the appropriate key (OPTION or ALT, COMMAND or CONTROL) *before* dragging it.

Figure 6.3 Example of Option–Dragging (Copy)

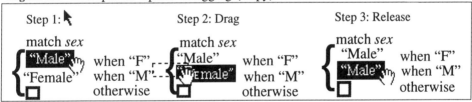

Changing a Formula

If you need to change a formula, double–click the corresponding column's selection area to bring up its Column Info dialog. To reopen the calculator window, click the picture of the formula in this dialog.

Note➡ You can take a shortcut to a column's formula window by option–clicking the selection area of the column. This eliminates using the Column Info dialog.

Deleting a function also deletes its arguments. Deleting a required argument or missing term from a function deletes the function as well. You can save complicated expressions that serve as arguments and paste them where needed. Use the **Copy** command to copy the arguments to the clipboard or scrapbook, or delete all but one function and argument and *peel* the function from the remaining argument as shown in **Figure 6.4**.

To peel a function from a single argument, first highlight the function with the mouse and then hold down the OPTION (or ALT) key and press the DELETE key. If you prefer to use the menus, you can choose **Clear** from the **Edit** menu in place of pressing the DELETE key. For example, to remove the Trim function (shown below) leaving only n, highlight the Trim(n) and press OPTION–DELETE on the Macintosh or ALT–DELETE under Windows.

Figure 6.4 Peeling an Argument

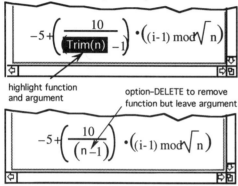

Or, you can use the hand to drag the argument on top of its function. See the section **Dragging Expressions** earlier in this chapter for a description of using the hand tool.

After you complete the changes above, the new values fill the column automatically when you click the Evaluate button or close the calculator window.

Once you have created a formula, you can change values in columns that are referenced by your formula. The calculator automatically recalculates all affected values in the formula's column.

Efficiency

Usually, it is not necessary to structure formulas with efficient evaluation in mind. Most formulas evaluate almost instantaneously regardless of their structure. This is because statistical functions and constant expressions are evaluated only once when a column's values are calculated.

However, when you are creating conditional expressions, keep in mind that **Match** evaluates faster and uses less memory than an equivalent **Condition**.

Consider the following two formulas for predicting a child's height from his age. In each case there is a base height of 63.0248015 inches to which a quantity is added depending on the value of the age variable.

$$63.0248015 + \begin{cases} \text{match } Age: \\ -4.8998015, & \text{when } 12 \\ -2.7390873, & \text{when } 13 \\ 1.14186507, & \text{when } 14 \\ 1.54662698, & \text{when } 15 \\ 1.30853174, & \text{when } 16 \\ 3.64186507, & \text{when } 17 \\ \bullet, & \text{otherwise} \end{cases} \qquad 63.0248015 + \begin{cases} -4.8998015, & \text{if } Age=12 \\ -2.7390873, & \text{if } Age=13 \\ 1.14186507, & \text{if } Age=14 \\ 1.54662698, & \text{if } Age=15 \\ 1.30853174, & \text{if } Age=16 \\ 3.64186507, & \text{if } Age=17 \\ \bullet, & \text{otherwise} \end{cases}$$

The Match conditional evaluates faster because the age variable is evaluated only once for each row in the data table. The If condition must evaluate the age variable at each if clause for each row until a clause evaluates as true.

Caution and Error Messages

Error and warning messages alert you that there is an error in your formula and warn you whenever you choose a command or use a JMP feature that can have unforeseen effects on calculated data. A beep often serves as the first warning of a problem, followed by a message if the problem persists. Calculator messages include both *Caution Alerts* and *Stop Alerts* as described below.

Caution Alerts
   allow you to continue a specific course of action but warn you of possible problems. They typically give you a choice of response buttons.

Stop Alerts
   detect errors that must be corrected before processing can continue. They often inform you that an invalid action has been ignored. Stop Alerts also appear each time you attempt to close the calculator window or click in another window while an invalid formula is in the formula display. When this happens click **OK** and correct or delete the invalid formula.

This section explains all alert messages that relate to the calculator window.

 Caution Alerts

The formula for Column 2 depends on Column 1. Do you want to remove Column 2's formula; or keep it, but change all references to Column 1 into missing?

This message appears when you attempt to delete a column in the data table that affects another column's values. In this example, the formula for *Column 2* includes a reference to *Column 1*. If *Column 1* is deleted and the formula is kept, all references to *Column 1* in the *Column 2* formula change to missing terms, which then calculate as missing values. If other columns depend on *Column 1*, this message will appear once for each column. If *Column 1* is deleted and the formula is not kept, the formula in *Column 2* will be removed and it will keep its calculated values. Click **Don't Delete** to keep *Column 1* and the dependent data in *Column 2*.

Are you sure you want to revert to the last formula saved?

This message appears when you click Revert in the Editing topic list of the function browser. Click **Yes** to discard all changes made to your formula since the last time you saved the data table. Click **No** to keep the current formula.

There is not enough available memory to undo this operation.

This message appears if you attempt to execute a command when there is insufficient memory for the **Undo** command. Click **Continue** to execute the command anyway. Click **Cancel** to ignore the command.

The formula for "Column 1" was created with a later version of JMP and may not be fully understood. Would you like to try and use it anyway?

This message appears if a formula is in a more recent format than the currently used version of JMP understands. You get this message when opening a data table created with a later version of JMP or when pasting a formula created with a later version of JMP into the Calculator. The new format may be perfectly understood, or there may be sections of it which are unintelligible. Click **OK** if you want to try to use the formula. Click **Cancel** to abort the operation and prevent the formula from being attached to the column.

Stop Alerts

The data type of the selected expression is *{numeric, character, row state}*, but it should be *{numeric, character, row state}*. Please change it to the correct type.

The selected (highlighted) expression evaluates to a data type that is invalid either in the context of the expression or the column type. Replace the highlighted expression with an expression that evaluates to the required data type as specified in the message. This message can appear any time the calculator attempts to evaluate a formula or when you reopen the data table.

Note → If the formula's column has the wrong data type, ignore the alert and change the data type using the Column Info dialog. After changing the data type, you may need to click the calculator's Evaluate button to refresh the column's values.

You must select a piece of the formula to manipulate first.

No expression in the formula display is highlighted. A function or feature of the calculator panel can affect a specific expression in the formula display only when that expression is highlighted.

You must choose a column from the column list first.

No column name is highlighted in the formula display. Some functions, such as subscript, require that the currently highlighted expression be a column name.

To add a clause, select the clause that you want to follow and press the comma key.

No clause or result expression is highlighted in the conditional expression. To add a clause to a conditional expression, first highlight all or part of the clause or result expression that is to precede the new clause. Then press the comma key or click ⌃.

To delete a clause, select it first. Then press the delete key.

No clause or result expression is highlighted in the conditional expression. To delete a clause from a conditional expression, first highlight all or part of the clause or result expression you want to delete. Then press the DELETE key or click 🔧.

There will always be at least one clause. You cannot delete it unless you delete the entire expression.

You tried to delete the only if, match or choose clause, or the otherwise clause. The JMP conditional expressions must have at least one if, match, or choose clause and an otherwise clause. Therefore, you must delete the entire conditional expression.

You can only have one non-missing argument in the selected expression. Cut or clear all of the arguments but one and try again.

There is more than one argument in the highlighted expression. If you need to delete a function but want to save one of its arguments, you can *peel* the function from the argument. To do this, change all the function's other arguments to missing terms by highlighting each argument and pressing the DELETE key. Peel the function by highlighting it in the formula and typing OPTION–DELETE. The above message appears if you attempt to peel a function from more than one nonmissing argument.

The selected column references a row that has not been evaluated yet. The subscript must be between 1 and i-1.

The formula references a row that is not yet evaluated. The calculator allows recursive formulas as long as they only reference a row number between 1 and i–1. Referencing any other row number is invalid. See the definition of Subscript in the Terms topic list in Chapter 5, "Calculator Functions," for more information about recursive formulas.

The selected column already has a subscript. To delete a subscript, just choose the column again.

Subscript (from the Terms list) has been chosen, but the highlighted column name is already subscripted. If you need to remove the subscript, highlight the column name in the formula display and choose the same column again from the column selector list. This replaces the subscripted column name with the same column name unsubscripted.

You can only use an index with the Sum (Σ) and Product (Π) functions.

Index has been chosen outside the scope of a summation or product function. Apply Index only within summations and products.

The formula for *Column 1* contains unrecognized expressions, probably because it was created with a later version of JMP. These expressions will be replaced with missing boxes.

A column formula contains expressions available only in a later version of JMP than you are currently using. All such expressions are replaced with missing terms.

Table Templates

JMP data table templates store formulas that are frequently applied to data. If you find yourself repeatedly entering the same formulas into different data tables, try creating a template table into which you paste new data. A *table template* is a table with either zero or one row and columns that store formulas. When you add or change rows of data, the formula columns automatically evaluate.

A collection of JMP table templates are saved as *stationery documents* in the SAMPLE DATA folder. When you open a stationery document, you do not open the document itself. The stationery document creates and opens a new JMP table that has the same contents as the stationery document. This protects you from inadvertently changing the table template. The following sections explain how to use table templates.

Grid Coordinates

The grid template provide a quick way to compute an evenly spaced x, y grid for three-dimensional plotting. There are two grid templates in the SAMPLE DATA folder:

- The template GRID COORDINATES (X=Y) produces a grid in which x and y both range from columns called **Start Grid** to **End Grid**.
- The template GRID COORDINATES (X≠Y) allows x to range from columns called **Start x** to **End x**, while y ranges from **Start y** to **End y**.

To use a grid template, first save it under a new name. Next, specify the grid endpoints in the first row of the new table. Then, enter a function of x and y into the Surface Formula column. You may want to paste a prediction formula from a response surface analysis into the Surface Formula. If you paste a formula from a different JMP table, use the option–click on the column name in the calculator to change the variable names to x and y in your formula.

At this point you still have only one row in your table. To obtain the grid, choose **Add Rows** from the **Rows** menu. Enter the number of rows you want minus 1 into the **# of Rows to add** box. As soon as you add rows, JMP evaluates all column formulas. To view the data in three dimensions, use the **Spin** platform.

Note➡ To create an n-by-n square grid, you add $n^2 - 1$ rows to the data table because there is already one row in the table.

If you need to alter the grid size or several axis ranges, remove all but the first row from your data table before doing so. This allows JMP to reevaluate formula columns for only the first row until all changes are made. See the COWBOY HAT and ODOR SURFACE data tables for examples that use grids.

Note ➡ Adding colors to your points based on the value of the column Surface Formula can enhance your spinning plot and is simple to do. Use the **Distribution of Y** to create a histogram of Surface Formula column. Get the hand tool and adjust the number of bins until there are at most twelve bins. Choose **Save Level Numbers** in the save ($) pop–up menu to save the bin levels into the data table. Now create a row state column and give it the formula

   hue(*Level Surface Formula*)

Actuarial Life Tables

Lifetime or survival data are often analyzed by computing actuarial life tables. Life tables estimate the survival, failure or probability, and hazard distribution functions of right-censored lifetime data. Because life tables are a series of formulas applied to a few columns of data, a table template is possible.

To use the LIFE TABLE TEMPLATE data table, save the table under a new name and then paste in your own data for the columns Years, # out, # lost, # dying, and Sample Size. The formula columns compute values for the survival, failure, and hazard distribution functions along with their standard deviations, and 95% confidence limits. Descriptions of each column are in the **Column Info** notes.

After you enter or paste your data into the life table template, you can graph the distribution functions using the **Plot** command in the **Summarize** menu.

See the ANGINA LIFE TABLE data table (Lee, 1980) for a life table example.

The Central Limit Theorem

The CENTRAL LIMIT THEOREM table template is used to demonstrate that the frequency distribution of the sample mean becomes normal as the sample size n increases. Columns of the table contain formulas to compute sample means for n equal to 1, 5, 10, 50 and 100. The samples are generated from a distribution of uniform random numbers raised to the fourth power (a highly skewed distribution). When you choose **Add Rows** from the **Rows** menu, each cell contains a sample mean. The number of samples taken is equal to the number of rows you add to the template.

To view the frequency distributions of the sample means, use the **Distribution of Y** platform to plot all columns. Add a normal curve to each histogram using the **Normal Curve** option in the check menu at the bottom of the report window. Note that the frequency distributions of large sample means follow a normal distribution more closely than the distributions for smaller sample means.

Nonlinear Modeling

The **Nonlinear Fit** command requires that you enter fitting and loss functions as formula columns in a data table. You may find it convenient to create templates for frequently used models. See Chapter 14, "Nonlinear Fitting," in the *JMP Statistics and Graphics Guide* for more information about the nonlinear table templates.

Keyboard Shortcuts

You can use the mouse or the keyboard shortcuts shown in **Table 6.2** to access calculator features.

Table 6.2 Keyboard Shortcuts

Replace a highlighted expression with the function or argument shown below by entering the corresponding key command.

?normal .... <Option>–?	Missing Element .... <Delete>
?uniform .... ?	n (Term) .... #
i (Term) .... @	π (Term) .... <Option>–p
Subscript .... [	Index ....]

Apply these functions to a highlighted expression by entering the key command.

((in the keypad) .... (	Inverse Tangent .... atan
) (in the keypad) ....)	Inv. Hyp. Cosine .... ahcos
x (in the keypad) .... *	Inv. Hyp. Sine .... ahsin
+ (in the keypad) .... +	Inv. Hyp. Tangent .... ahtan
– (in the keypad) .... –	Mean .... _
÷ (in the keypad) .... /	Natural Log .... ln
± (in the keypad) .... <Option>––	Not .... !
And .... &	Or .... |
Combination, New .... <Option>–\	Power .... ^
Common Log .... log	Quantile .... %
Condition, If .... {	Radical .... <Option>–v
Cosine .... cos	Std Deviation .... <Option>–j
Exponential .... exp	x < y .... <
Delete .... <delete>	x = y .... =
Hyperbolic Cosine .... hcos	x > y .... >
Hyperbolic Sine .... hsin	x ≠ y .... <Option>–=
Hyperbolic Tangent .... htan	x ≤ y .... <Option>–,
Insert .... ,	x ≥ y .... <Option>–.
Inverse Cosine .... acos	Σ .... <Option>–w
Inverse Sine .... asin	Π .... <Shift–Option>–p

Use <Shift> with keyboard shortcuts that includes a shift–key character.

Other keyboard equivalents include the following key actions:

- Press the ENTER key or the RETURN key to enter a focused item from the calculator panel in a formula. For more information about calculator focus, see the section called **Working with Formulas** in this chapter.
- Up and down arrow keys traverse the column selector list and function browser list.
- Left and right arrow keys switch panels of the function browser list.

Chapter 7
Report Windows

This chapter describes features common to all report windows in JMP. These include standard in window controls and special additional JMP features.

To make the most of each analysis, all report windows have button controls, pop-up menus, and help windows. In addition, you can select from eight tools that change the function of the mouse. This puts report features literally at your fingertips.

You can customize the appearance of report windows by changing type styles, adding colors and labels to plots, and more. Once you've explored the report windows on your monitor, you can print them directly from JMP. If you want to include analyses in your own comprehensive report, you can journal report windows to word-processing files or copy any part of a report window into another application.

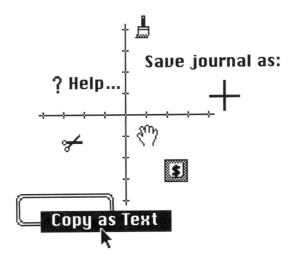

Chapter 7
Contents

Report Windows 196
   Standard Window Features 196
   Border Menu Options 197
   Tear–Off Menus 198
   JMP Tools 199
   Reveal Buttons 200
   Pop–up Menus 201
   Report Fonts and Sizes 201
   Changing the Plot Background 202
   Formatting Report Table Columns 202
Graphical Displays 202
   Resizing Plots and Graphs 203
   Customizing Axes 203
   Highlighting Points in Plots and Using Row States 205
   Zooming 207
   Using the Annotate Tool 207
Help Windows 208
   Help from the About JMP Screen 209
   Help From the JMP Statistical Guide 210
   Help from Buttons in Dialogs 210
   Help from a Platform Window 210
   Help by Clicking an Item 211
   Additional Help commands under Windows 212
Copying from JMP Report Windows 213
   Copying Graphical Displays 213
   Copying Text Only 214
   Copying a Text Report to a JMP Table 214

Pasting Between Applications 215
Journaling Reports 215
Printing Reports and Journals 217

Report Windows

All analysis platforms produce report windows. These windows contain both text reports and graphical displays. They are designed to be convenient in several ways:

- Although report windows consist of complex plots and calculations, they appear on the screen almost instantaneously most of the time.
- Individual plots can be resized, and text reports can be concealed to optimize the use of space on the monitor.
- Report windows always print each individual plot and text report completely on one page if possible.
- Many areas of report windows have contact sensitive–help.

All or part of any report window can be copied to other applications, journaled to a special file, or printed directly from JMP.

Standard Window Features

JMP reports are displayed in standard windows with scroll bars, a resize box, a zoom box, and a close box. Most JMP report windows also have other special buttons and pop–up menus like those illustrated in **Figure 7.1** and discussed in the following sections.

Figure 7.1 Report Window

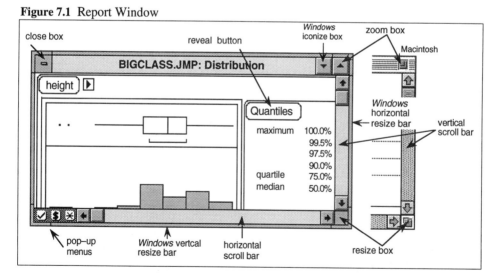

Border Menu Options

There are three special icons to the left of the horizontal scroll bar in all JMP report windows. Each icon displays a pop–up menu showing options for that particular display:

The check mark pop–up menu lists options that are different for each report window. For example, in the Distribution of Y analysis, this menu has options for fitting a normal curve to each histogram, for switching displays from vertical to horizontal orientation, and more.

The dollar sign identifies a pop–up menu for saving data generated by the current analysis. The new data appear as additional columns in the data table used to create the report. These commands differ for each analysis platform. For example, on two platforms you can save predicted values and on another you can save ranks.

The asterisk accesses the pop–up menu shown in **Figure 7.2** that lists the following options for tables and graphical reports:

- **Help** shows a help window for the current platform. The help windows have buttons you can use to show further help.
- **Reveal All** opens all text reports in the active report window.
- **Conceal All** closes all the reveal buttons, concealing all text reports in the active report window.
- **Title** and **Footnote** commands focus editable areas at the top and bottom of report windows. When you select these commands, you see the text area outlined and a size box in its lower right corner. The text areas also appear when you click near title or footnote text.

  You can size the text areas to include as many lines of text as you want. You can also select the **Title** and **Footnote** commands multiple times and create independent text entry areas. To delete title or footnote areas, delete the text.
- **Alignment** displays a pop–up menu that lets you position the text relative to the size of the text area.
- **Font Type** displays a pop–up menu that lets you select the same font attributes you used as preferences for text, heading, or display title. See the section called **Report Fonts and Sizes** later in this chapter for a discussion of preferences.

  Note➔ If you option–click the title or footnote text areas when they are not focused for editing, the pop–up menus shown to the right in **Figure 7.2** display for aligning text and selecting a font type.

  Figure 7.2 Option–click in Text Area to Enter Title and Footnote

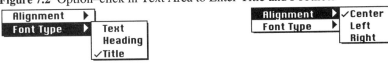

- **Print Preview** imposes page size outlines on reports to show how printing tables, plots, and graphs will be arranged on the printed page.

Tear-Off Menus

The following situations can involve the need to select items frequently from menus or palettes:

- alternating between the arrow tool and the brush, scissors or other selection from the **Tools** menu
- assigning colors to multiple groups
- assigning markers to multiple groups.

To help speed up the process of tool, color, and marker selection, the **Tools** menu and the color and markers pallettes are *tear-off* menus. This means you can click and drag them from the main menu to a convenient place on your desktop. They remain open until you close them, and you select from them by clicking the selection you want. When you start a new session, JMP remembers where the floating menus were and they appear where you left them.

Tear-off menus remain visible on the surface of tables and platforms. Their selections are dim unless the appropriate JMP conditions exist to use them. For example, the special tools are enabled only when an **Analyze** or **Graph** platform is the active window. The colors and markers are enabled only when there are highlighted data table rows.

Figure 7.3 Tear-off Menus Give Floating Tools and Markers Palettes

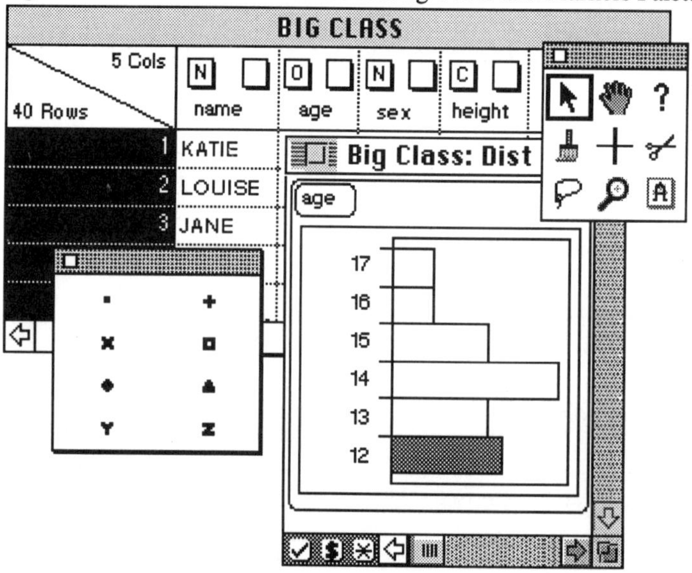

JMP Tools

To expand the capabilities of the mouse in report windows, you can choose special mouse cursors from the **Tools** menu. Mouse actions have different effects depending on which tool you choose. The six tools are described briefly in the list that follows: Chapter 2, "The Menu Bar," describes these tools in detail.

For convenience you can *tear–off* the **Tools** menu and place it anywhere on your desktop.

▶ The arrow is the default tool and the standard cursor. It is the most versatile tool. For more information about the arrow, see the section **Highlighting Points in Plots**.

🖐 The hand tool (or *grabber*) directly manipulates histograms, spinning plots, and the scatterplot matrix. It works like the arrow on all other parts of report displays.

? The question mark tool accesses help windows. Click and hold the mouse button on an area within a report window if you need information. The help window remains on the screen until you release the mouse button. When you shift–click with the question mark, the help window remains on the screen until you close it.

🖌 The brush tool highlights points in plots. The brush tool works like the arrow when it is not in a graphical display. For more information about the brush, see the section called **Using the Brush** later in this chapter.

+ The crosshair tool measures the axes position of points on a plot. When you drag the mouse around on a plot, you see crosshairs that follow the movement of the mouse. The values at the intersection with the X and Y axes display as you drag the crosshair.

✂ The scissors tool selects an area in the report window to copy.

⌒ The lasso tool lets you highlight an irregular area of points in plots. Drag the lasso around any set of points. When you release the lasso it automatically closes and highlights the points within the enclosed area. Use SHIFT-lasso to drag the lasso around discontiguous irregular areas of points.

🔍 The magnifying glass lets you automatically zoom in on a plot area of interest. Option–clicking always displays the plot in its original form. Examples of zooming with the magnifier are shown later in this chapter.

A The annotate tool creates a simple text area on any graphical display. You can key in simple notes or use it to enhance a graphical display. There are examples shown later in this chapter.

Reveal Buttons

The Buttons shown in **Figure 7.4** are called *reveal* buttons. Click a reveal button to alternately reveal and conceal its corresponding text report.

Figure 7.4 A Reveal Button

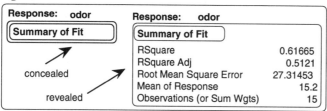

Sometimes reveal reports are nested in layers as shown in **Figure 7.5**. You can either click to open each layer individually or SHIFT–click to open all layers. SHIFT–clicking again conceals all the layers of nested reports.

Figure 7.5 Nested Text Reports

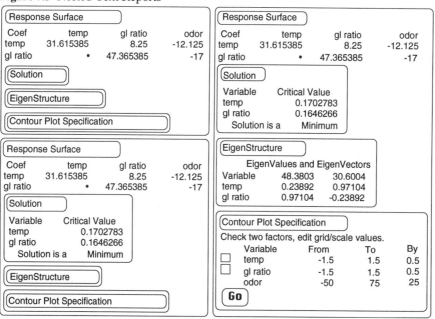

You can also reveal or conceal all text reports using the **Reveal All** and **Conceal All** commands listed in the asterisk pop–up menu. This menu is at the left of the horizontal scroll bar in each report window. See the section called **Border Menus** earlier in this chapter for a description of each asterisk menu option.

Pop-up Menus

▶ Many report windows include one or more pop-up menus (accessed by the icon shown at left). These pop-up menus list different options and commands depending on the analysis platform.

Report Fonts and Sizes

One way you can customize the appearance of reports is to change text styles with the **Preferences** command in the **File** menu. The Preferences dialog, shown in **Figure 7.6**, has pop-up menus called **Font** and **Size** on the Macintosh (**Fonts...** under windows). The **Font** menu lists different type faces, and the **Size** menu lists both point sizes and styles. You can change the fonts for text, headings, and titles:

- Text refers to the body of reports.
- Headings refers to column headings.
- Titles refers to window titles and reveal buttons.

On the Macintosh, if you hold down the OPTION key while selecting from the **Font** menu, JMP assigns the font you choose to all three report parts simultaneously.

Figure 7.6 The Preferences Dialog

Macintosh Preferences Dialog

Windows Preferences Dialog

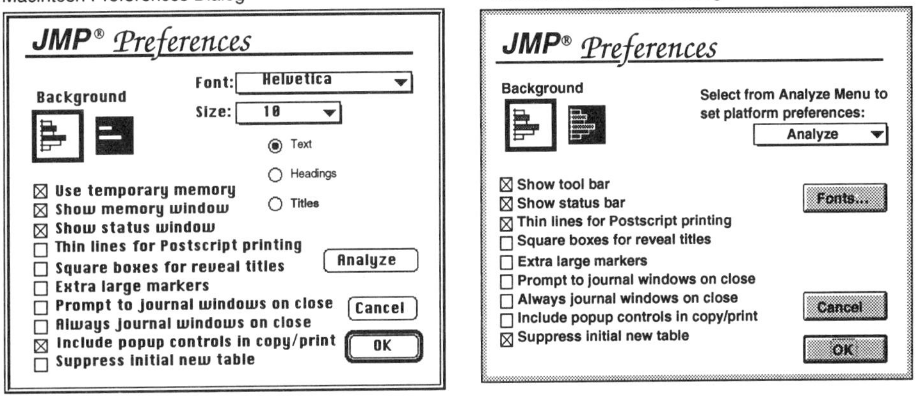

Note➔ The Preferences **Analyze** button displays a dialog listing the **Analyze** menu commands. When you click an **Analyze** command, a second dialog lists the platform's text reports and graphical displays. To tailor the initial display of an analysis platform, select the options you want, and then click **Save**. Whenever you select an **Analyze** command, the window components you saved as preferences automatically display. The other preferences are described in Chapter 2, "The Menu Bar."

Changing the Plot Background

Graphs can have either white or black backgrounds. To set the background, choose the **Preferences** command from the **File** menu and click either the black or white background icon (see **Figure 7.6**). The background preference does not affect all plots. For example, the spin plot has its own background setting.

Formatting Report Table Columns

You can format columns of numbers in tables using the dialog shown in **Figure 7.7**. Double-click anywhere in a column to see the dialog. Edit the **Width** and **Decimal** boxes with the format you want. When you click **OK** the entire column formats accordingly.

By default, JMP uses what it considers the best format (denoted *Best* in the Column Info dialog) for numbers both in statistical reports and in the data table. Best format displays as many decimals as the field width of the number allows. It also truncates trailing zeros.

Figure 7.7 Formatting Table Columns

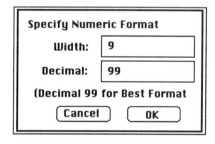

Graphical Displays

Graphical displays are plots of data within JMP report windows. Special options are available to manipulate these plots in each analysis window.

Resizing Plots and Graphs

To resize a plot or graph, first click so that a small box, called the *resize box*, appears in the lower right corner of the plot frame. When you drag this box, the height and width of all plots in that frame adjust independently of other frames in the same report window (see **Figure 7.8**). The plot frame resize box adjusts widths of plots in the same frame independently but adjusts the heights of all plots in one frame as a single unit.

Figure 7.8 Plot Frame Resize Boxes

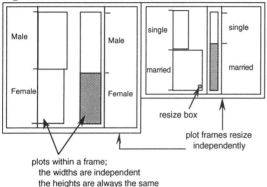

When you use this special resize box, the option, shift, and command keys have the following special effects:

OPTION–resize on the Macintosh
ALT–resize under Windows

   adjusts a plot in 8–pixel increments.

SHIFT–resize

   adjusts the plot frame perfectly square.

OPTION–shift–resize on the Macintosh
ALT–shift–resize under Windows

   adjusts the plot frame both perfectly square and in 8–pixel increments.

COMMAND–resize on the Macintosh
CONTROL–resize under Windows

   in any resize box adjusts all like plots in that window simultaneously. For example, you can have scatterplots and mosaic plots in the same window. If you use COMMAND–resize (CONTROL–resize) in one scatterplot resize box, all scatterplots resize together. The mosaic plots stay the same.

Customizing Axes

You can double–click in the numeric axis areas of most plots or charts to see an axis customization dialog. Customization features on the dialog depend on the data type of the axis and with the specific platform. **Figure 7.9** shows the dialog for the Overlay platform, which supports the grid option, reference lines, and log axes.

Figure 7.9 The Axis Customization Dialog

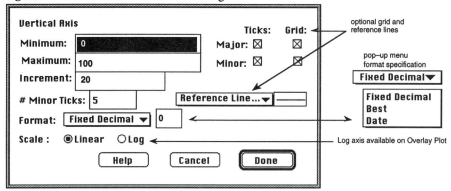

The example in **Figure 7.10** illustrates customization of both numeric axes given by the Fit Y by X platform:

- The axes of both plots have minor tick marks added to the horizontal and vertical axes.
- Both the horizontal and vertical grid lines on the major tick marks better define the geography of a densely packed cluster of points in the first plot.

The plot on the right is an enlargement or *zoom* of the point cluster that shows between 0 and 500 in the plot to the left. The enlarged plot is obtained by reassigning the maximum and minimum axis values and changing the grid lines specification.

Figure 7.10 Enlarging a Plot Section

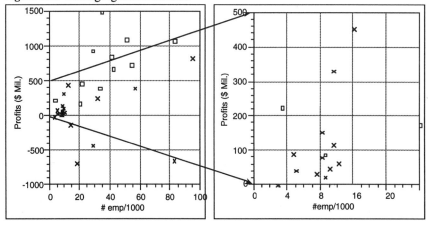

If your axis is a JMP date value, the **Format** pop-up menu gives you options for the date display and an additional pop-up lists date increments for tick marks. JMP dates are numeric values that correspond to the number of seconds since January 1, 1904. JMP dates and the date display options **Short**, **Long**, and **Abbrev** are described in the section **Characteristics of Data** in Chapter 3, "JMP Data Tables."

Figure 7.11 shows the axis specification to plot monthly data for 12 months with major tick marks every month and one minor tick between them. Selecting a date interval from the date increment pop-up menu divides the JMP date (number of seconds) into the appropriate units to give the plot scale you want for your data.

Figure 7.11 Format Axis with Dates

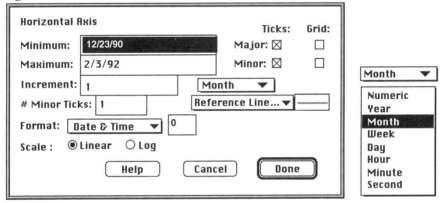

Highlighting Points in Plots and Using Row States

Most graphs display columns from the data table. Points in scatterplots have different appearances depending on the active row state characteristics. Row state conditions include select (highlight), exclude, hide, label, marker, and color. There are 7 JMP markers and 64 JMP colors (available on color monitors). Row states are described in Chapter 3, "JMP Data Tables."

Assigning Labels

Label is a special row state that assigns a label to one or more rows in the data table. Row numbers are the default labels when there is no column assigned as a label column..

To designate a label for every row using values in a column, assign the role of *label* to one column in the spreadsheet using the column's role assignment box.

To apply a row state, select rows in any representation of the data table and choose the row state you want from the **Rows** menu. The data table and plots automatically show the effect of the row state.

Saving Row States

You can save row states permanently in a row state column. The two pop-up menus at the top of a row state column activate or save row state information. The **Copy to RowState** or **Add to RowState** command activates row state information. **Copy from RowState** and **Add from RowState** commands copy the active row state information into the row state column. Active row states save when you close the table and are active when you open it.

Highlighting Points

Clicking a point in a plot with the arrow cursor highlights (selects) the point and the corresponding row in the current data table. The corresponding row's label appears on the display and persists as long as you hold down the mouse button. If you option–click a point, then its label is copied to the clipboard.

When you click the arrow in a display, previously selected data are no longer selected. You can either select a new point or no points depending on where you click. To extend a selection, hold down the shift key while you click new points.

The brush tool in JMP highlights a group of points delineated by a rectangle. As you drag the brush over a plot, points that fall within the rectangle are temporarily highlighted. Since highlighted points are larger, the brush tool is like a magnifying lens as it passes over points. When you release the mouse button, the highlighted points within the rectangle remain selected.

You can shift–drag the brush tool to extend the current selection. The new selection includes the previous selection and all points that pass within the rectangle while the SHIFT key is pressed. The extended selection areas need not be contiguous.

You can also OPTION–drag (ALT–drag under windows) the brush tool to resize its rectangular selection area. This feature lets you drag a rectangle around the area containing all the points you want to select. You can also use OPTION–drag to form a thin vertical or horizontal rectangle. This shape acts like a slicing tool that can traverse and highlight across either axis.

If you hold down the command key and drag the brush tool, the selection status of points within the rectangle continuously inverts. This causes the points to blink on all representations of the data.

If you COMMAND–OPTION–drag on the Macintosh (CONTROL–ALT–drag under Windows) with the brush tool and then let go (*(push* the rectangle), the rectangle bounces around in the frame on its own. The speed and direction you push (or drag) the mouse determines the speed and direction of the bouncing rectangle. Click the mouse button in the report window to stop the rectangle from bouncing.

The lasso tool lets you highlight an irregular area of points in plots. To highlight points, drag the lasso to around any set of points. When you release the lasso it automatically closes and highlights the points within the enclosed area. Use SHIFT-lasso to drag the lasso around discontiguous irregular areas of points.

Zooming

The magnifying glass tool lets you automatically zoom in on any area of a plot. When you click the magnifier, the point or area where you click becomes the center of a new view of the data. The plot enlarges approximately 25%, focusing on the area in to give you a closer look at interesting points or patterns. When you drag the magnifier to form a rectangular area, this area enlarges and occupies the whole axis.

OPTION–click on the Macintosh or ALT–click under Windows at any time anywhere in the plot frame to restore the original plot.

The POLLEN data table (pollen.jmp) in the SAMPLE DATA folder dramatically illustrates the magnifier. The data were devised with a hidden pattern (message) buried in over 3,000 points. The Fit Y by X platform displays the whole table as the dense cloud shown to the left in **Figure 7.12**. Clicking the magnifier twice produces the plot in the middle, and clicking twice more reveals the message showing in the plot on the right.

Figure 7.12 Example of Zooming with the Magnifier Tool

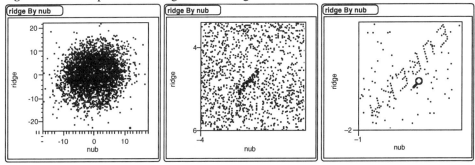

Using the Annotate Tool

 The annotate tool creates a editable note area whereever you click in a JMP display window. You can key in notes and remove them at a later time, or use the annotate tool to enhance a JMP graphical display.

To use the Annotate tool, select it from the Tools menu and click or drag a rectangle in a display window. A small white editable text box appears. The annotation can be manipulated in the following ways:

- When you click outside the text box it becomes yellow and is no longer editable, and the size of the box automatically adjusts to fit the size of the text you entered.
- Double–click inside the note and it again becomes editable.
- Click near the middle area of the text box and drag to move it around.
- Click inside the box to see handles in the middle of each edge, and a size box in the lower right corner.
- Drag from side handles in any direction to draw a line. The lines stays in place and moves with the text box after you release the mouse. To remove a line click its origin.
- To remove a text box, simply drag it out of the window.

The example in **Figure 7.13** is a window with the graphical results of the **Ternary** command in the **Graph** window. The palette from the **Tools** menu has been torn off and placed in the window next to the ternary plot. TheAnnotate tool created the two notes on the platform that give special information about the Tools palette and the Ternary plot.

Figure 7.13 Example of Annotated Display window

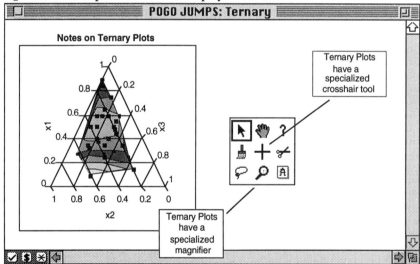

Help Windows

Help windows give information about questions you have while using JMP. Sometimes they include pictures and examples, but they always include textual explanations. Many help windows also have buttons to access further help.

The file called JMP Help, which is distributed with JMP, contains the help information. Be sure to keep it in the same folder as JMP.

You can access help windows almost anywhere in JMP:

- The question mark tool shows automatically when you drag the cursor over the buttons in the About JMP window. When you click these buttons, you have access to help for the whole JMP application.
- If you click the **Guide** button in the About JMP window, you have access to a guide of statistical methods.
- Many dialogs have their own help buttons.
- Help for each platform window is accessed by the asterisk pop–up menu located to the left of its horizontal scroll bar.
- Help for specific items in a JMP window show when you select the question mark from the **Tools** menu and click the item.

Help from the About JMP Screen

If you have general questions or want to see a list of broad help topics, choose **About JMP** from the Apple menu on the Macintosh. Under Windows, choose the **Contents** command in the **Help** menu. This displays the JMP window shown in **Figure 7.14**, which is the topmost help window. This main JMP help window has buttons for further help. All the help documents are arranged hierarchically with buttons for accessing more detailed information.

Figure 7.14 About JMP Window

Help From the JMP Statistical Guide

The first button in the JMP window is your online statistical navigation guide. To see a list of statistical and graphical procedures available in JMP, click the **Guide** button at the top of the About JMP window. Under Windows you can access the Statistical Guide by choosing the **Statistical Guide** command in the **Help** menu on the main menu bar.

The Statistical Guide is a scrolling statistical index. When you click an analysis in the guide, the directions on how to do the analysis appear beneath the index, as shown in **Figure 7.15**. The directions tell you the menu, command and options to use for the analysis or topic you selected.

Figure 7.15 Statistical Guide

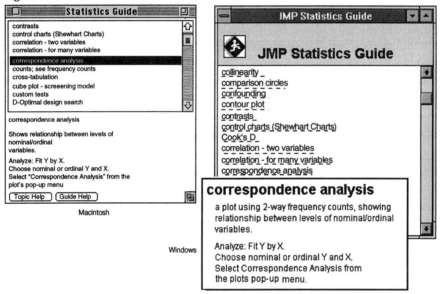

Help from Buttons in Dialogs

Most dialog windows in JMP include a **Help** button. For example, there are help buttons on the Specify Model dialog, the calculator window, the Preferences dialog, and on data manipulation dialogs.

Help from a Platform Window

If you want help using an analysis platform, select **Help** from the asterisk pop–up menu located to the left of the horizontal scroll bar.

For example, **Figure 7.16** shows help for the **Distribution of Y** platform with a second layer of help for the histogram component of the platform.

Figure 7.16 Help from a Platform Window

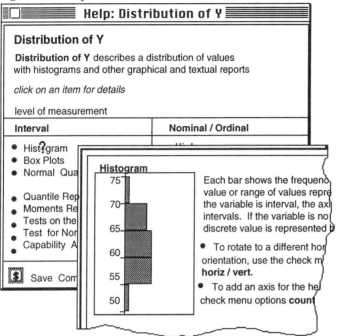

Help by Clicking an Item

When you need help concerning a specific item in a JMP report window, select the question mark from the **Tools** menu and click the item. For example, when you click **Std Error** (see **Figure 7.17**), a help screen displays. The help screen persists as long as the help tool is on top of it. Also, you can shift–click to open the help screen as a window that stays open until you close it with the close box.

Figure 7.17 Clicking an Item for Help

Parameter Estimates				Help: Std Error
Term	Estimate	Std Error		
Intercept	-26.92308	12.7495		**Std Error**
temp	-12.125	9.38337		is the standard error, an estimate of the standard deviation
gl ratio	-17	9.38337		of the distribution of the parameter estimate. It is used to
temp*temp	31.615385	13.771		construct t tests and confidence intervals for the parameter.
gl ratio*temp	8.25	13.2701		
gl ratio*gl ratio	47.365385	13.771		

If a help window has buttons for further help as shown in **Figure 7.18**, it persists on the screen until you close it.

Figure 7.18 Help Window with Additional Help Buttons

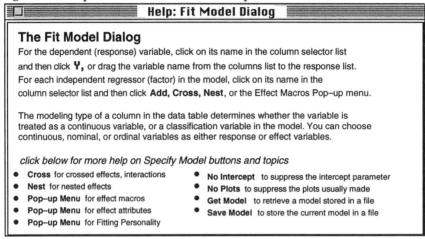

Additional Help commands under Windows

The Microsoft Windows **Help** menu as additonal commands called **Search for Help On...** and **How to Use Help**. The **Search for Help On...** displays the scrolling list of topics like that shown in Figure **Figure 7.19**. When you select a topic and click Go To, you immediately see the topic–specific help.

Figure 7.19 Help Window with Additional Help Buttons

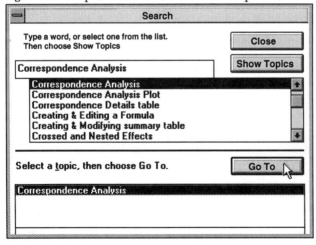

Copying from JMP Report Windows

You can use the standard copy and paste operations to move data and graphical displays from JMP to other applications. Although the **Edit** menu includes both **Cut** and **Copy** commands, they perform the same tasks in document windows. **Cut** copies all or part of the active report window into the clipboard but does not delete the image from JMP. There is also a special command called **Copy Report** (discussed later in this chapter) that copies only the text from an active report window.

Copying Graphical Displays

On the Macintosh, when you copy from a report window, the image is stored in the clipboard as a standard Macintosh picture. By default you copy the entire report window. You can copy all of a report window or part of any report window. The scissors tool scrolls when you drag it outside a report window.

If you want to copy part of a report window, select the scissors tool from the **Tools** menu. Drag the scissors diagonally to draw a rectangle over the area you want to copy. The tip of the upper blade marks the starting and ending points of the selection. As you drag, a rectangle appears and selects the area to copy as shown to the left in **Figure 7.19**.

If you want the scissors tool to select specific components of a report, hold down the OPTION key (ALT key uner windows) and click or drag. Each component of the report highlights as a separate unit as shown by the right-hand plot in **Figure 7.19**. Hold down both the option and shift keys to extend a selection of discrete components.

Figure 7.19 Selecting Plot Areas with the Scissors Tool

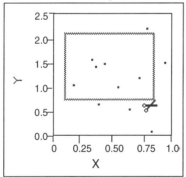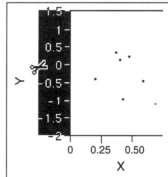

You can change the selection if necessary. Drag the scissors over a new area or extend the current selection by shift-dragging the scissors across an additional area. Extensions do not have to be contiguous, and you can add as many extensions as you want. To deselect all areas, click the report window (without dragging).

Copying Text Only

To copy text only from a report window, choose **Copy as Text** from the **Edit** menu as shown in **Figure 7.21**. This command copies all text from the active report window, ignoring any current selection made with the scissors tool. Each field within a text report is separated by tabs so that you can edit the text easily in a word processing spreadsheet or application.

Figure 7.21 The Copy as Text Command

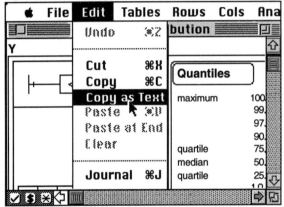

Copying a Text Report to a JMP Table

You can copy a JMP text report into a JMP data table by selecting it with the scissors tool, using the Copy as Text command, and then using Option–Paste at End to copy it to a new empty JMP table.

If you hold down the OPTION key (ALT key under windows) with **Copy as Text** (OPTION–**Copy as Text**) and again with **Paste at End** (OPTION–**Paste at End**), the values in the first line of information on the clipboard are used as column headers. The remaining information is the table content. This lets you copy a table from a JMP analysis window or other application and paste it into a new data table with appropriate column names.

For example, the table to the left in **Figure 7.10** is the result of a JMP analysis. The scissors tool selects the table and OPTION–**Copy as Text** copies the table to the clipboard. Then, OPTION–**Paste at End** when a new data table is active produces the table shown to the right.

Figure 7.22 Parameter Estimates Table Copied to a JMP Data Table

Parameter Estimates	
Term	Estimate
Intercept	66.5
HBARS[down-up]	.5
DYNAMO[off-on]	-6
SEATS[down-up]	1.75
TIRES[hard-soft]	-1.25
GEAR[low-medium]	-11.25
RAINCOAT[off-on]	0.25
BRKFAST[no-yes]	0.5

Untitled 1		
2 Cols	N	C
8 Rows	Term	Estimate
1	Intercept	66.5
2	HBARS[down-up]	0.5
3	DYNAMO[off-on]	-6
4	SEAT[down-up]	1.75
5	TIRES[hard-soft]	-1.25
6	GEAR[low-medium]	-11.25
7	RAINCOAT[off-on]	0.25
8	BRKFAST[no-yes]	0.5

Pasting Between Applications

If you have plenty of memory, you can run JMP and other applications simultaneously to copy and paste between them. Otherwise, you can paste JMP images into the scrapbook desk accessory for intermediate storage.

JMP copies data in PICT data format with hierarchical grouping information (used by applications such as MacDraw). However, there are some differences between the PICT data and the on–screen image. For example, points in JMP plots save as very small squares in order to improve their printed appearance.

Not all applications interpret PICT data in the same way. For example, some drawing programs ignore clipping information (information about invisible parts of pictures). In such programs, some lines appear longer than they appeared in JMP. For example, fitted lines and confidence curves may extend beyond axis boundaries when copied from JMP and pasted into another application.

See the section called **Printing Reports** for information about copying JMP PICT data into other applications and printing.

Journaling Reports

Journaling is a special feature of JMP that helps you save analysis results in a word processing document. When you select the **Journal** command from the **Edit** menu, all text and pictures from the active JMP analysis window are written to the journal document. If an area of an analysis window is selected with the scissors tool, the **Journal** command saves the selected area as a picture instead of saving the entire window.

The JMP journal window opens after you first select the **Journal** command. It displays within JMP as illustrated in **Figure 7.24**. The JMP journal window is an on-screen representation of the document you will see when you open it with a word processor. JMP appends text and graphs from the active analysis window to the end of the journal window each time you again select the **Journal** command.

Figure 7.24 A Journal Window

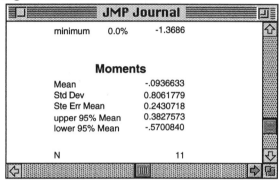

The journaled text and graphs cannot be changed until you close the file and open it with a word processor. However, when a JMP journal is open you can

- append text and graphs from the active analysis window to the end of the journal window each time you again select the **Journal** command
- type text notes at the bottom of the journal window
- scroll through the journal window
- copy individual graphs and text reports from a JMP analysis window and paste them at the end of a journal window
- copy text and pictures from other Macintosh applications and paste them at the end of a journal window
- print the contents of the journal window at any time using the **Print** command in the **File** menu
- hide the journal window by selecting **Hide** from the **Window** menu.

When you close the journal window (using the close box or **Close** command), a save dialog like the one shown in **Figure 7.23** prompts you for a disk location, file format, and name for the JMP journal. You select the file format you want from the pop-up men u showing beneath the **Save Journal as:** prompt. When you click **Save** the journal is saved using the Claris XTND system of file translators. After the journal is saved, you open it with a word processor of you choice for preparing reports.

Note➡ You cannot open an existing journal document from within JMP. After you close your journal it has the file format you specified. It is not a JMP document. If you want continued access to a journal from within JMP but want it out of the way, leave it open and use the **Hide** command in the **Window** menu.

Figure 7.23 The Save Journal Dialog

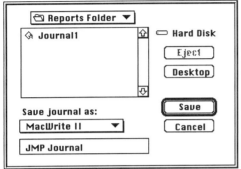

Printing Reports and Journals

Select **Print** from the **File** menu to print revealed text reports and plots from the active report window, or print an open journal window. Reports print best on PostScript output devices such as the LaserWriter, but they also print on Quickdraw laser printers and ImageWriters.

When you print a JMP text report or graph, the text reports and graphical displays automatically rearrange on each printed page for best fit. Some controls and icons do not appear on the printouts. You can use the **Include Pop–up Controls in Copy/Print** preference to see pop–up controls. The **Print Preview** selection in the Star (*) border menu draws page–size rectangles on the report window to show you how printing will occur.

The journal prints as it appears on your screen.

The PICT files that store graphical displays generated by JMP contain special PostScript (Adobe Systems Corp.) information to make them print better. With this special information, lines and curves are thin, and dashed lines have uniformly sized segments. When you copy a JMP picture into most word processing applications, this special information is retained. However, some drawing programs discard extra PostScript commands.

The following windows cannot be printed:
- the model specification dialog
- the statistical guide.

Appendix A
Questions and Answers

Q? What is the difference between JMP, and JMP SERVE?

A: JMP SERVE is a specially tailored package of multiple JMP licenses. JMP SERVE is used in Mac labs, where a number of Macintoshes are connected to a server. You can purchase a JMP master disk with a counter that allows a given number of concurrent JMP users. This reduces the unit cost per user. JMP SERVE comes with documentation for the number of the JMP–Pack you buy.

Q? Can I read my JMP version 1 and Version 2 data tables with JMP version 3?

A: Version 3 of JMP reads previous Version files and converts them to Version 3 files. If you want to save your previous copy, use **SAVE AS** to create a new Verson 3 file.

Q? What do I do when I run out of memory?

A: If you are in the middle of a session, close any windows that you don't need. When you close windows, it frees memory both for the data and for the the program code used by the window.

On the Macintosh, if you quit of JMP, you can reset its *Application Memory Size*, which is the memory JMP tries to get when it is launched. From the Finder, select the JMP application file and choose the Get Info command under the **File** menu. This brings up a window of information about the JMP file. In the lower right corner you see a field labeled *Application Memory Size (K)*, in which you can type the amount of memory you think you may need. We distribute JMP with that field set to 2000K, but recommend you set it to several times that value if you have enough memory.

Q? I have a SAS data set. How do I get it into a JMP data table?

A: Methods for exchanging data between SAS and JMP vary with the version of SAS you are using. These methods are described in the Hypercard stack, called SAS Transport Help, in the JMP folder.

Q? What is the difference between the **Paste** and **Paste at End** commands?

A: The **Paste** command requires rows or columns (or both) to be selected and pastes data from the clipboard into the region defined by the selection, writing over any values that are there.

The **Paste at End** command uses selected rows or selected columns *but not both*. If there are selected rows, **Paste at End** creates new columns and pastes the data into the new columns at the end of the selected rows. If there are selected columns, **Paste at**

End creates new rows and pastes the data into the new rows at the end of the selected columns. The **Paste at End** command never writes over values in the spreadsheet.

Note: If there are no selected rows or columns, **Paste at End** copies the data from the clipboard to the end of the spreadsheet. It behaves as if all the columns are selected. To transfer data from another application into JMP, copy it from the application to the clipboard. Then use the **New** command from the **File** menu to create a new empty spreadsheet. When you select **Paste at End**, the contents of the clipboard are written to the new table with columns and rows created as needed.

Q? **What do the options on the SAS Transport dialog mean?**

A: A SAS transport file can contain multiple members of a SAS data library. JMP cannot tell the number of members nor the name and size of each member until the entire transport file is read. By default, JMP reads the entire SAS transport file and creates a spreadsheet window for each member. If the transport file is large and has many members, this can be very memory demanding! The **SAS Transport** import options let you save the members as JMP files rather than opening spreadsheet windows for them, or select only the one that you want to open as JMP spreadsheet windows.

Q? **What is the header information used for that is exported by the Save As command when I click on Text format and the Header option?**

A: These options write a JMP data table in text format with JMP header information preceding the data lines. The header contains information about the JMP data table, as well as information about the columns. The following is an example of a header extracted from a JMP export file. The header is followed by the data values arranged in rows and columns as they are organized in the JMP data table.

```
Header Begin
    TableName="big class"
    NRows=45
    NColumns=5
    Mode=ReadOnly
    TableNotes="this is a test"
    Column Begin
        Label="name"
        DataType=Character
        FieldWidth=0
        Measure=Nominal
        Notes="Student's Name"
    End
            .
            .   more column header information
            .
    Column Begin
        Label="height"
        DataType=Numeric
        Format=Best12.
        Measure=Interval
```

            End
End
ALFRED M 14 69.0 112.5
ALICE   F 13 56.5   84.0
BARBARA F 13 65.3   98.0
    .
    .
    .
        more data lines

Since the header is just a set of information written by JMP at the beginning of a standard text file, it can easily be created with any text editor. The format of JMP export header is fairly flexible. There can be multiple items on a line. The header can be as complete as the one shown above created by JMP, or as brief as

```
Header NRows = 5
```

The keyword `Header` is required to identify the beginning of header information. The keywords `Begin` and `End` are used both for the main header and for the column information to bracket more than one item in a list. Text strings with embedded blanks should be enclosed in double quotes ("). Keywords have to be spelled exactly as they appear in the JMP export header.

Table B.1 and **Table B.2** summarize the keywords and the column information used as JMP header information.

Table B.1 Summary of JMP Header Keywords

Header	Start of the JMP Export Header
`TableName=`	data table name
`NRows`	maximum number of rows to be read
`NColumn=`	maximum number of columns
`Mode=`	the word `ReadOnly` if the data table is locked; omit this information otherwise
`TableNotes=`	data table notes which appear in the **Notes** field of the **Table Info** dialog
`TColumn`	start of column definition

Table B.2 Summary of Items inside a Column Definition List

Key Word	Information
`Label`	a string of up to 31 characters for column name
`DataType=`	one of the words Numeric, Character, RowState, which define the data type of the column
`FieldWidth=`	maximum number of characters for a character column

`Format=`	format specification for a numeric column. usually in the form n.d, where n is the number of display position for the number, and d is the number of decimal places; the special key word **Best** can precede the format specification as in the case of the variable Height in the above example.
`Measure=`	the measurement level of the column
`Mode=`	the keyword `Lock` if the column is locked; omit this information otherwise
`Notes=`	column notes which appear in the Notes field of the **Col Info** dialog

This text header is very useful for importing data. When importing data that has no JMP header information (the usual situation), JMP must make some basic assumptions about the data which may or may not reflect the intended usage. For example, JMP interprets a column of integer data as numeric when importing. Should the data happen to be row state data, JMP still reads it as simple numeric data. If you define the data type of the column in the header of the text file as RowState, JMP interprets the column of data as row state data instead of numeric data.

Another useful example is to add the `NRows=` header information at the beginning of a data file and specify the maximum number of rows for JMP to import. This gives you the ability to limit the number of rows that JMP reads from the text file.

Q? Why don't my statistical tests and estimates agree with GLM?

A: You probably have an ordinal classification variable. In some kinds of analyses JMP treats ordinal variables differently than nominal variables but SAS treats them the same. See Appendix A in the *JMP Statistics and Graphics User's Guide* for a comparison of SAS and JMP.

Appendix B
What's New in Version 3 of JMP?

General

In most role assignment dialogs you can drag variables from any list to any other list in addition to selecting a variable and clicking the role button you want.

The variable list boxes in most dialogs show the modeling type of each variable. A small icon with C, N, or O shows to the left of the variable names. Other dialogs have been simplified and there is a more uniform arrangement of button actions.

You can *tear off* the **Tools** menu, and the colors and markers palettes in the **Rows** menu.

There is a new main menu called **Graph** that lists the following commands: **Bar/Pie Charts, Overlay Plot, Spinning Plot, Pareto Charts, Control Charts,** and lists the new plotting commands **Contour Plot** and **Ternary Plot**.

The **Analyze** menu has new commands (**Cluster** and **Survival**) and enhancements to existing commands. The **Spin** command is the **Spinning Plot** command in the **Graph** menu.

The **Specify Model** command name is changed to **Fit Model.**

The **Analyze** menu has new commands (**Cluster** and **Survival**) and enhancements to existing commands.

The axis customization dialog has an additional feature that lets you specify reference lines on all platforms that support grid lines.

Design of Experiments

The Version 2 JMP *Design* product is integrated into JMP Version 3. You access the Design of Experiments module with the **Design Experiment** command in the **Tables** menu.

The Choose Design Type dialog is organized in a more logical order, and there is a D–Optimal Design button that gives help and refers you to the Fit Model dialog.

The Choose Design dialog includes a General Factorial selection that builds a table of all combinations of levels for variables you specify.

The design you select automatically produces a JMP table

Most models have a Make Model option that opens the Fit Model dialog with an appropriate model for the selected design.

Main Menus

The File Menu
The **Import** command has a pop-up menu that lists **Text** and **SAS Transport**. If you have ClearAccess® installed on your machine, it also lists **Run ClearAccess Script**, **Edit ClearAccess Script**, and **Launch ClearAccess**.

There are additional preferences that let you choose thin line for postscript printing, square reveal button boxes, large markers, automatic journaling, and show pop-up controls for copying and printing.

The Edit Menu
The commands are rearranged in a more logical sequence.

The **Copy Report** command name is changed to **Copy as Text**.

There are new commands called **Find** and **Find Next** for finding character strings in a data table.

The Tables Menu
The commands are rearranged in a more logical sequence.

The new **Split Columns** command splits a single column into several columns according to the number of levels in a specified Row ID column or combinations of levels in multiple Row ID columns. The **Split Col** command performs the inverse operation of the existing **Stack Cols** command.

The new **Attributes** command generates an *Attributes table* from the active data table, called its *Source table*. You can update the column attributes of a data table by editing its contents table and using an **Update Source** command.

The new **Design Experiment** command accesses JMP's Design-of-Experiments module.

When you hold down the SHIFT key and choose the **Subset** command, the rows of the subset remain linked to the original table. When you highlight a row in the subset table, the corresponding row highlights in the original table. You can use this technique as often as you want and each new subset is linked to all other tables in which it exists.

The Rows Menu
The new **Move Rows** command displays a dialog to move any set of selected rows to any row location in the table.

The colors palette has 65 colors.

The **Select** submenu has an additional command called **Where** that lets you select rows based on values or ranges of values in selected columns.

The **Color/Marker by Col** command uses the levels of the variable you select to color or mark points in plots.

The Cols Menu

The new **Add Columns** command displays a dialog for adding multiple columns of the same type at any column location in the data table.

The new **Move Columns** command displays a dialog for moving multiple columns to any location in the data table.

The New Column dialog has radio buttons offering data validation options that restrict a column's values to a list or a range of values, and lets you name the columns with the same prefix.

The Analyze menu

The **Analyze** menu has new commands (**Cluster** and **Survival**) and enhancements to existing commands.

Distribution of Y
- You can rescale the histogram and normal quantile plot axes of continuous variables
- There is an outlier box plot as well as a quantile box plot for continuous variables.
- There is a Uniform option, which causes the axes for all numeric variables to be scaled the same.
- There is a Uniform option, which causes the axes for all numeric variables to be scaled the same.
- The **Smooth Curve** option fits a smooth curve to the histogram of each continuous variable using nonparametric density estimation. It displays with a slider, which determines the range of Y values used to determine curve estimates.

Fit Y by X: regression platform has added options to the fitting pop–up menu.
- A paired t–test presents standard computations and an innovative graphical display
- The check (✓) pop–up menu has an additional option called **Smooth Curve**. This option fits a smooth curve to the histogram of each continuous variable using nonparametric density estimation. The smooth curve has with a range slider that lets you set the kernel standard deviation of Y that defines the range used to determine curve estimates.

Fit Y by X: one–way Anova platform has two pop–up menus called Analysis and Display that show beneath the plot.
- The Analysis pop–up Quantiles, Means-Anova/t-test, Means-Std Dev-Std Err, all the Compare and Nonparametric selections as before, UnEqual Variances, and Set Alpha Level. The plots appear automatically but can be toggled off with options in the Display pop–up menu.
- The Display pop–up lists Show Points, Quantile Boxes, Means Diamonds, Means Dots w Error Bars, Std Dev Lines, Comparison Circles, Connect Means, and Proportional Axis as display options.

Fit Y by X's and **Fit Model** have been consolidated with no loss of functionality. The **Fit Y by X's** command was a special case of **Fit Model** when a model has no complex effects (has only main effects or simple regressors). The only difference between the commands was the dialog in which you specifed a model.

Fit Model has a new dialog giving the platform extended functionality.
- An **Effects** pop–up menu lets you designate any highlighted effect as a random effect, variance effect, or response surface effect.
- The **Effects Macros** pop–up menu lets you automatically specify the effects model as a full factorial, a factorial to a degree, a sorted factorial with effects listed in order of their degree, a response surface, a mixture response surface, or a polynomial.
- A fitting pop–up menu with selections Standard Least Squares, Screening, Stepwise, Manova, Loglinear Variance, Nominal Logistic, Ordinal Logistic, Proportional Hazard, and D–Optimal Design.
- The standard least squares platform has new options that perform Inverse Prediction, A Durbin–Watson test, and Correlation of Estimates, and a save option to save the formula for the prediction formula for the standard error.
- For response surface models, you can create a QuickTime movie of two variables changing over levels of a third variable.
- The Manova selection has an automatic repeated measures selection.

Nonlinear Fit
- Loss functions can now have parameters.
- Confidence Limits are computed differently.
- There is a check box to use if you are doing maximum likelihood and want the proper scaling of standard errors and confidence intervals.

Spin is in the **Graph** menu and is called **Spinning Plot**.

Correlations of Y's is a new name for the **Y's by Y's** command. There are the new platform features: pairwise correlations, nonparametric correlations, and principal components/factor analysis. You can also do principal components in the **Spinning Plot** command in the **Graph** menu but **Correlations of Y's** is better if you have many variables.

Cluster clusters rows of a JMP table either hierarchically or using a k–means method. The hierarchical clustering method displays results as a dendogram followed by a plot of the distances between clusters. The five clustering methods available are Average Linkage, Centroid, Ward's minimum variance, Single linkage, and Complete Linkage. You can also request a k–means clustering approach for larger data tables.

Survival command lets you choose product–limit (Kaplan–Meier) life table statistics, parameteric regression modeling, and Cox proportional hazard modeling. The product–limit life table computes statistics for single or multiple strata. The platform displays an overlay plot of estimated survival function for each strata and for the whole sample. Optionally, it displays exponential and Weibull failure plots to graphically check whether those distributions are appropriate for further regression modeling. The parameters for these distributions are estimated. It also computes the log rank and Wilcoxon statistics to test homogeneity of strata, and can compute competing risks when there are multiple causes of failure.

The parametric and Proportional Hazard modeling selections display the Fit Model dialg tailored for thespecific kind of analysis.

The Graph menu

Bar/Pie Charts has display options to hide the options buttons so they don't appear in journals or printed reports and to request Vertical legends when the overlay option is in effect. Also, you can request **Show Points** without the **Connect Line** option in effect. The result is a point chart. Previously, if you did not have the **Connect Line** option in effect, the **Show Points** option was dimmed.

Overlay Plots has a step option to produce step plots, and a range option that connects the minimum and maximum points at each X value when the overlay option is in effect

Spinning Plot is the new name in the **Graph** menu for the **Spin** command that was previously in the **Analyze** menu. You can record QuickTime movies of spining plots.

Control Charts offers the new charts UWMA, EWMA, Cusum charts, and can create and save a periodogram.

Contour Plot produces a contour plot from rectangular or nonrectangular data. It also lets you create and save a grid of points with estimated response points.

Ternary Plot constructs a plot using triangular coordinates.with an optional contour of the surface produced by the sample data. The ternary contour has the same capabilities as the Contour plot platform.

The Tools menu

The **Tools** menu (and the colors and markers palettes in the **Rows** menu) can be torn off and placed anywhere on your desktop.

The scissors tool scrolls when dragged outside a window.

The crosshair tool displays the changing values at the X and Y axis intersection as you click and move (drag) it about the plot.

There are three new tools:
- The Magnifier tool lets you zoom repeatedly on a plot. Double clicking restores the plot to its original scaling.
- The Annotate tool gives you the ability to placd simple text messages anywhere in a JMP report windwo. The text boxes are movable, resizable, and you can draw a line from the box to an item of interest on the plot.
- The Lasso tool lets you select irregular sets of points in plots .

The Data Table

The data type handles an unlimited number of rows, restricted only by the amount of memory available.

The column attribute previously referred to as "measurement level" is now called "modeling type."

The modeling type previously called "interval" is now called "continuous."

New cursor forms show on cells in columns that have values restricted to a list of values or a range of values.(see **Cols** menu).

Formulas are preserved when you subset.

The row state column pop–up menu has new commands that adds values to and from the active status. The **Add to RowState** and **Add from RowState** commands preserve existing row state information. The **Copy to RowState** and **Copy from RowState** commands do not preserve existing row state information.

The surface organization of the data table is new and more compact.

You can select rows or find values in cells with new commands in the **Edit** and **Rows** menus (see **Edit** and **Rows** menus sections).

When you hold down the option key and select the Subset command in the Tables menu, the subset is linked to the original table. When you highlight rows in the subset, the corresponding rows higlight in the original table.

The Calculator

If you copy a formula containing parameters to another calculator, the parameters are matched by name with existing parameters. If they don't exist, new parameters are automatically created. However, you may have to set initial values.

You can save JMP 3.0 data tables in JMP 2.0 format with the **Save As** command in the **File menu.**

Additional character functions include Subset, Contains, Word and Item.

The new Variables function category lets you define temporary variables for a single column formula.

The Transcendental functions include Gamma, Incomplete Gamma, Log Gamma and Squash functions.

The Numeric functions include a Round function.

The Row State functions include a This Row function, which was available in late Version 2 but not documented.

Scripting

JMP 3.0 supports AppleScript. If you have AppleScript, use the Script Editor's "Open Dictionary" command to view JMP's AppleScript dictionary. Documentation is included as a separate document.

New or Changed Key Board Shortcuts for Menu Commands (Macintosh)

Command	Menu	Previous Key	New Key
Import	File	⌘I	none
Find	Edit	none	⌘F
Find Next	Edit	none	⌘G
Find	Edit	none	⌘F
Find Next	Edit	none	⌘G
Replace	Edit	none	⌘R
Replace & Find Next	Edit	none	⌘H
Locate Next	Rows	⌘A	⌘+
Locate Previous	Rows	⌘B	⌘–
Select All (rows)	**Rows**, Select pop–up menu	none	⌘A
Select Where (rows)	**Rows**, Select pop–up menu	none	⌘?
Move to First	Cols	⌘F	⌘<
Move to Last	Cols	⌘L	⌘>
Column Info	Cols	none	⌘I
Redraw	Window	⌘R	⌘D

Index

A
abbreviated date format 90
About JMP command (Help menu) 68
actuarial life table template 189
Add Columns command (Cols menu) 47, 77
Add from RowState command 93
Add Rows command (Rows menu) 40, 75
Add to RowState command 93
adding and deleting columns (data table) 76
adding and deleting rows (data table) 75
ALT key (with Copy and Paste commands) 84
Analysis platform preferences 15
Analyze menu commands 50-55
    Cluster 54
    Correlation of Y's 53
    Distribution of Y 50
    Fit Model 52
    Fit Y by X 50-51
    Nonlinear Fit 53
    Survival 55
annotate tool (Tools menu) 63, 199, 207
appending data tables 30, 114-116
argument (calculator formula) 136
Arrange Icons command (Window menu) 65
arrow cursor 72
Assign Roles command (Cols menu) 42
assignment function (calculator) 154
Attributes command (Tables menu) 33, 78, 97, 125
attributes table 33, 78, 125-128
    adding, moving, deleting source table columns 78, 128
    columns of 78, 126-128
    options 128
    updating a source table 128

B
Bar/Pie charts (Graph menu) 56
best numeric format 46, 89
brush tool (Tools menu) 62, 199, 206
By Mode (Group/Summary command) 24

C
calculator 131-192
    caution alerts 185
    character functions 149
    clause insert and delete buttons 138
    clauses and conditions 154-158
    comparison functions 153
    components of 135
    constant entry field 138
    date and time functions 168-169
    defining parameters 173
    derivatives 175
    dragging expressions 182-184
    editing functions 175
    error messages 185-187
    examples 131, 133-135
    formula display area 139
    function browser 138, 141
    functions 143-176
    keyboard short cuts 192
    keypad functions 140-141
    numeric functions 144
    probability functions 160-162
    random number functions 159
    row state functions 170-172

statistical functions 163-167
stop alerts 187
temporary variables 174
terminology 136-137
terms functions 143
transcendental functions 146
window 135-136
work panel 138
Cartesian join (data tables) 123-124
Cascade command (Window menu) 65
cell focus (data table) 73
char to num function (calculator) 152
character data 44
character functions (calculator) 149-152
choose function (calculator) 157
clause (formulas) 138
clauses and conditions (calculator) 154
Clear All Roles command (Cols menu) 43
Clear command (Edit menu) 20, 82
Clear Row States command (Rows menu) 39
ClearAccess (Import command) 12, 94
Close type Windows (Window menu) 65
Cluster (Analyze menu) 54
color function (calculator) 172
Color/Marker by Col command (Rows menu) 40
colorOf function (calculator) 172
Colors command (Rows menu) 38, 91
Cols menu commands 42-49
    Add Columns 47, 77
    Assign Roles 42
    Clear All Roles 43
    Column Info 49
    Delete Columns 49, 78
    Hide Columns 48
    Move Columns 47
    Move to First 47
    Move to Last 47
    New Column 43-46, 76
    Unhide 48
Column Info command (Cols menu) 49
columns (data table) 42-49
    adding, creating 34, 47, 77
    attributes 33, 44-46
    best numeric format 89
    calculating values 45, 131-176, 177-192
    character 79
    Cols menu 42-49
    creating 43
    data source 44, 45
    data type 44, 88
    data validation 44
    date and time formats 46
    deleting 49, 78
    fixed decimal format 89
    formats 89-90
    hiding and unhiding 48
    modeling type 45, 88
    moving 47
    names of 71
    New Column command 43-46, 76
    number of 71
    numeric 79
    selecting 73-74
comparison operators (calculator) 153
computing column values 131-176, 177-192
Concatenate command (Tables menu) 30, 97, 114-116
concatenating data tables
    different column names 115-116
    same column names 114
concatentate characters (calculator) 151
constant entry field (calculator) 138
contact sensitive help 211, 212
contains function (calculator) 151
Contents command (Help menu) 66
continuous data 45, 89
Contour Plot (Analyze menu) 60
Control Charts (Graph menu) 59
Copy as Text command (Edit menu) 19, 214
Copy command (Edit menu) 19, 82, 213
Copy from RowState command 93
Copy to RowState command 92
Correlation of Y's (Analyze menu) 53

count function (calculator) 145
cross cursor 72
crosshair tool (Tools menu) 63, 199
current row 74
cursor forms in data tables 72
customizing plot and chart axes 203
Cut command (Edit menu) 19, 82

D

data
    calculating values 44, 131-176, 177-192
    character 44
    continuous 45, 89
    cutting and pasting 82-83
    dates and times 46, 89
    finding values 21, 79
    formats 46, 89-90
    importing 80-82
    journaling 94
    modeling type 45, 88, 127
    nominal 45, 89
    numeric 44, 89
    ordinal 845, 9
    replacing values 79
    row states 44, 89
    source 44-45
    types of 44, 88, 127
    validation 44
data tables 69-129
    Add Columns command 47
    adding columns 34, 43, 47
    adding rows 40, 75
    attributes table 33, 78, 125
    cell focus 73
    Close command 13
    Cols menu 42-49
    column characteristics 127
    computing column values 131-176, 177-192
    concatenating 30, 114-116
    creating a data table 75-85
    creating columns 43-46, 76

current row 73, 74
cursor forms 72
deleting columns 34, 49
deleting rows 41, 75
elements of 71
entering and editing data 78-85
excluding rows 37
exporting data 87
generating grid coordinates 188
Group/Summary command 24-26
hiding columns 48
hiding rows 37
Import command 10-13
importing data 80-82
joining 31-33, 117-124
labeling rows 37
life table template (actuarial) 189
locating rows 40
moving columns 47
moving rows 40
New Column command 43-46
New command 9, 75
Open command 9
Print command 17
Revert command 15
Rows menu 36-41
Save As command 13
Save command 13
searching for text 21
selecting rows and columns 38, 73-74
sorting data 27, 107
Source table 24, 33
splitting columns 28-29, 110-111
stacking columns 27, 108-110
Subset command 26
Summary table 24
Table Info command 33
Tables menu 23-35
transposing 30, 111-113
updating 128
variable roles 42
data validation 84

date and time formats 46, 89, 90
date and time functions (calculator) 168
dates (JMP dates, SAS dates, formats) 90
Delete columns command (Cols menu) 49, 78
Delete rows command (Rows menu) 41, 75
derivatives (calculator) 175
deselecting rows and columns 75
Design Experiment command (Tables menu) 35, 97, 129
distribution functions (calculator) 160
Distribution of Y (Analyze menu) 50
double cross cursor 72

E

Edit menu commands 18-22
    Clear 20, 82
    Copy 19, 82, 213
    Copy as Text 19, 214
    Cut 19, 82
    Journal 20, 94, 215-217
    Paste 19, 82, 215
    Paste at End 20, 82, 213
    Search 21-22, 79
    Undo 18, 82
editing functions (calculator) 175
educational services 242
Exclude/Include command (Rows menu) 37, 91
Exit command (File menu) 17
exporting data 13, 80, 87
expression (calculator) 137

F

File menu commands 9-17
    ClearAccess (Import) 12, 94
    Close 13
    Exit 17
    Import 10-13, 80-82
    New 9, 75
    Open 9
    Page Setup 17
    Preferences 15, 201
    Print command 17

    Quit 17
    Revert 15
    Save 13
    Save As 13, 87
finding text 21-22, 79
Fit Model (Analyze menu) 52
Fit Y by X (Analyze menu) 50-51
fixed decimal format 46, 89
font and font size preferences 201
formats (data) 46, 89-90
formatting report table columns 202
formulas 131, 133, 179-184
    argument 136
    calculator components 137
    changing 183
    character functions 149-152
    clause 138
    clauses and conditions
    comparison functions 153
    conditions functions 154
    cutting and pasting 181
    date and time functions 168-169
    derivatives 175
    dragging expressions 182-184
    editing functions 175
    efficiency 184
    examples 131, 133-135
    expression 137
    focused work area 180
    function 137
    function browser 141
    keyboard short cuts 192
    keypad functions 140-141
    means function 165
    missing term 137
    missing value 137
    Of( ) function 166
    order of precedence 179-180
    parameters in 173
    probability functions 160-162
    product function 166
    quantiles function 164

random number functions 159
row state functions 170-172
selecting expressions 181
standard deviation function 165
statistical functions 163-167
summation function 166
temporary variables 174
term 137
terminology 136-137
terms functions 143
transcendental functions 146
function browser (calculator) 141
functions (calculator) 137, 144-176

G

gamma function (calculator) 148
Graph menu commands 56-61
    Bar/Pie charts 56
    Contour Plot 60
    Control Charts 59
    Overlay Plots 57
    Pareto Charts 58
    Spinning Plot 58
    Ternary Plot 61
grid lines on plots 204
Group/Summary command (Tables menu) 24-26, 73, 97, 98-105
    By Mode 24
    grouping dialog 99-100
    source table 24
    subgroup option 25, 105
    subset analysis 101
    summary statistics 24-25, 98, 102-105
    summary table 24, 98-101, 105

H

hand tool (Tools menu) 62, 199
help 66-68, 208-212
    About JMP screen 209, 210
    buttons in dialogs 209, 210
    help tool (?) for contact sensitive help 211, 212
    report windows pop-up menu 210
    statistical guide 209, 210
Help menu commands 66-68, 212
    About JMP 68
    Contents 66
    How to Use Help 68
    Search for Help On 66
    Statistical Guide 67
help tool (Tools menu) 63, 199
Hide Columns command (Cols menu) 48
Hide command (Window menu) 65
Hide/Unhide command (Rows menu) 37, 91
highlighting points in plots 206
How to Use Help (Help menu) 68
hue function (calculator) 171

I

if, otherwise function (calculator) 155
Import command (File menu) 10-13, 80-82
importing data 11-13, 80-82
    ClearAccess facility 12
    SAS transport files 11
    text file dialog 11
importing SAS data sets 12
incomplete gamma function (calculator) 148
insert and delete buttons (calculator) 138
Invert Selection command (Rows menu) 39
item function (calculator) 151
I-beam cursor 72
Join command (Tables menu) 31-33, 97, 117-124

J

joining data tables 31-33, 117-124
    by matching columns 31-33, 120
    by row number 31
    Cartesian join 31, 123-124
    drop multiples option 32, 122
    dropping or keeping columns 118-120
    include non-matches option 132, 22
    join by row number 117-118
    join dialog 31
Journal command (Edit menu) 20, 93, 215-217
journaling JMP results 20, 215

K

key pad (calculator) 138
keyboard arrows (spreadsheet) 86
keyboard shortcuts (calculator) 192

L

Label/Unlabel command (Rows menu) 37, 91
lasso tool (Tools menu) 62, 199, 206
length function (calculator) 152
life tables (actuarial) 189
list check cursor 72, 84
list check option (entering data) 84
Locate Next command (Rows menu) 40
Locate Previous command (Rows menu) 40
locating selected row 74
log axis scale 203
log functions (calculator)
logical functions (calculator) 157
long date format 90
lowercase function (calculator) 152

M

magnifier tool (Tools menu) 63, 199, 207
marker function (calculator) 172
markerOf function (calculator) 172
Markers command (Rows menu) 38, 92
match function (calculator) 156
max and min functions (calculator) 144
means function (calculator) 165
menu bar 7-22
    Analyze menu 50-55
    Cols menu 42-49
    Edit menu 18-22
    File menu 9-17
    Graph menu 56-61
    Help menu 66-68
    Rows menu 36-41
    Tables menu 23-35, 95-129
    Tools menu 62-63
    Window menu 64-65
missing term (calculator) 137
missing value (calculator) 137
mod function (calculator) 144

modeling type (data) 45, 88-89
Move Columns command (Cols menu) 47
Move Rows command (Rows menu) 40
Move to Back (Window menu) 65
Move to First command (Cols menu) 47
Move to Last command (Cols menu) 47
munger function (calculator) 149

N

New Column command (Cols menu) 43-46, 76
    list checking 84
    range checking 84
    types of data 79, 89
    validation of data 84
New command (data table) 9, 75
New Data View (Window menu) 65
nominal data 45, 89
Nonlinear Fit command (Analyze menu) 53
nonlinear modeling 173
nonlinear modeling (table templates) 190
normal random number function 159
num to char function (calculator) 152
number of rows function (calculator)
numeric data 44, 89
numeric functions (calculator) 144

O

Of( ) function (calculator) 166
Open command (File menu) 9
open cross cursor 72
OPTION key (Copy and Paste command) 84
ordinal data 45, 89
Overlay Plots (Graph menu) 57

P

Page Setup command (File) 17
parameters (calculator) 173
Pareto Charts (Graph menu) 58
Paste at End command (Edit menu) 20, 82, 213-214
Paste command (Edit menu) 19, 82, 215
pasting between applications 215
Preferences command (File menu) 15-17, 201
Print command (File menu) 17

Print Preview option 197, 217
printing report windows 217
probability functions (calculator) 160-162
product function (calculator) 166

Q

quantiles function (calculator) 164
Quit command (File menu) 17

R

random number functions (formulas) 159
range check cursor 72, 84
range check option (entering data) 84
Redraw command (Window menu) 65
reference lines on plots 204
report windows 193-217
    annotating 207
    assigning labels to points 205
    asterisk (*) pop-up menu 197
    check mark (✓) pop-up menu 197
    contact sensitive help 211
    Copy as Text command 214
    Copy command 213
    copy text to a JMP table 214
    customizing axes 203
    font and size preferences 201
    footnotes 197
    formatting axes 201
    formatting tables 202
    grid lines 203
    highlighting points 205, 206
    JMP Tools 62, 199
    log axis scale 203
    pop-up menus 201
    preferences 15, 197
    Print Preview option 197, 217
    printing 217
    resizing plots and graphs 203
    reveal buttons 200
    save ($) pop-up menu 197
    titles 197
    using row states 205
    zooming 207

resizing plots and graphs 203
reveal buttons 200
Revert command (File menu) 15
round function (calculator) 144
row number function (calculator) 170
row state data 89
row State functions (calculator) 170-172
row state data 44
row states 91-93
    add to and from 92
    Colors command 91
    copy to and from 92
    Exclude/Include command 37, 91
    functions (calculator) 170-172
    Hide/Unhide command 37, 91
    Label/Unlabel command 37, 91
    markers 38
    Markers command 91
    saving 205
    selection 38, 92
rows (data table)
    adding rows 40, 75
    deleting rows 41, 75
    number of 71
    row states 37-38, 91-93
    selecting 73-74
Rows menu commands 41
    Add Rows 40, 75
    Clear Row States 39
    Color/Marker by Col 40
    Colors 38, 91
    Delete Rows 41, 75
    Exclude/Include 37, 91
    Hide/Unhide 37, 91
    Invert Selection 39
    Label/Unlabel 37, 91
    Locate Next 40
    Locate Previous 40
    Markers 38, 92
    Move Rows 40
    Select 73, 93
    Select submenu 38

S

SAS transport files 11, 80-82
Save As command (File menu) 13, 87
Save command (File menu) 13
save commands (analysis reports) 78
Save SAS Transport file 13
scaling axes 204
scissors tool (Tools menu) 63, 199
Search command (Edit menu) 21-22, 79
Search for Help On (Help menu) 66
Select command (Rows menu) 73, 93
Select submenu (Rows menu) 38, 74, 92
selecting rows and columns 18, 73-74
selection status (row state) 92
Set Window Name (Window menu) 65
shade function (calculator) 171
short date format 90
Show Colors Palette (Window menu) 64
Show Markers Palette (Window menu) 64
Show Tools Palette (Window menu) 64
shuffle (random number function) 159
Sort command (Tables menu) 27, 107
sorting a data table 107
source table
    Attributes command 125
    Group/Summary command 98
Spinning Plot (Analyze menu) 58
Split Columns command (Tables menu) 28, 97, 110-111
squash function (calculator) 148
Stack Columns command (Tables menu) 27, 97, 108-110
stacking data table columns 108-110
standard deviation function (calculator) 165
starting JMP 71
statistical functions (calculator) 163-167
statistical guide (help) 67, 209, 210
subgroup option (Group/Summary command) 26
subscript function (calculator) 143
Subset command (Tables menu) 26, 97, 106
substring function (calculator) 151
summary statistics (Group/Summary command) 24-25, 98, 102-105
summary table 73, 98-101
    Group/Summary command 73, 98-101
    source table 98
    subset analysis 101
    summary statistics 105
    transposing source table groups 113
summation function (calculator) 166
Survival (Analyze menu) 55

T

Table Info command (Tables menu) 33, 97, 125
table templates
    actuarial life table 189
    central limit theorem 190
    grid coordinates 188
    nonlinear modeling 190
Tables menu commands 23-35, 95-129
    Attributes 33, 77, 78, 125-128
    Concatenate 30, 77, 114-116
    Design Experiment 35, 77, 129
    Group/Summary 24-26, 73, 77, 98-105
    Join 33, 77, 117-124
    Sort 27, 77, 107
    Split Columns 28-29, 77, 110-111
    Stack Columns 27, 77, 108-110
    Subset 26, 77, 106
    Table Info 33, 77, 125
    Transpose 30, 77, 111-113
tear off menus 198
technical support 242
temporary variables (calculator) 174
term (calculator) 137
terms functions (calculator) 143
Ternary Plot (Graph menu) 61
tick marks on axes 204
Tile command (Window menu) 65
time formats 46, 89
time functions (calculator) 168
titles and footnotes 197
Tools menu commands 62-63, 199

    annotate tool 63, 199, 207
    brush tool 62, 199, 206
    crosshair tool 63, 199
    hand tool 62, 199
    help tool 63, 199
    lasso tool 62, 199, 206
    magnifier 207
    magnifier tool 63, 199
    scissors tool 63, 199
Tools menu commands 62-63
transcendental functions (calculator) 146
Transpose command (Tables menu) 30, 97, 111-113
transposing a data table 111-113
    by groups 112
    simple transpose 111
    with a label 112
trigonometric functions (calculator) 146
trim function (calculator) 151

U

Undo command (Edit menu) 19, 82
Unhide command (Cols menu) 48
uniform random number function 159
uppercase function (calculator) 152

V

validation of data (range and list checks) 84

W

What's New in JMP Version 223-229
Window menu commands 64-65
    Arrange Icons 65
    Cascade 65
    Close type Windows 65
    Hide 65
    Move to Back 65
    New Data View 65
    Redraw 65
    Set Window Name 65
    Show Colors Palette 64
    Show Markers Palette 64
    Show Tools Palette 64
    Tile 65

word function (calculator) 151

Z

zooming (magnifier tool) 207

SAS Institute Services

When you need Help...

If you are a registered user, technical support is as near as your telephone. To register, send in the registration card enclosed with the JMP product. As a registered JMP User, SAS Institute provides you with unlimited Technical Support for one year.

The Technical Support department is staffed by a group of product experts committed to providing you with knowledgeable and timely support. Technical support is available by phone (919-677-8008), fax (919-677-8123) AppleLink (SAS.TECH), or mail, Monday through Friday, 9 AM to 5 PM Eastern Time.

When you want to learn more...

Look into instructor–based training of JMP software offered by SAS Institute Inc. Seeing your data values can help you understand them better and make discoveries you might miss looking only at numbers. Two courses on JMP software are available:

- **Interactive Data Analysis Using JMP Software**

This two–day course is for scientists, researchers, engineers, quality technicians, instructors, and others who want to analyze data using JMP. In this course you learn how to analyze a variety of data using interactive methods. The statistical visualization capabilities of JMP unite classical statistical methods with new techniques in exploratory data analysis and statistical graphics.

Before selecting this course, you should have completed an undergraduate statistics course, be familiar with basic descriptive statistics, and techniques such as simple linear regression and one–way analysis of variance. You should d also know how to use a mouse to select, click, shift–click, drag, and how to select commands from menus.

- **Design and Analysis of Experiments Using JMP Software**

This one–day course is for quality technicians especially in manufacturing. It is valuable for scientists and researchers responsible for designing experiments, conducting them, and analyzing results.

Before selecting this course, you should have completed Interactive Data Analysis Using JMP software or have equivalent knowledge and experience. You should also have completed a graduate–level course in experimental design or have equivalent knowledge and experience.

When and Where

The JMP courses are taught as back–to–back in SAS regional training centers, and can also be taught at your location. To register for a public course, call a course registrar in Cary, NC at 919-677-8000, extension 5005, or call the Institute training center nearest you. For more information about the content of the courses, call 919-677-8000, extension 7205. To schedule a course at your location, call an Education Account Representative, at extension 7321.